Classical Relativistic Electrodynamics

Advanced Texts in Physics

This program of advanced texts covers a broad spectrum of topics which are of current and emerging interest in physics. Each book provides a comprehensive and yet accessible introduction to a field at the forefront of modern research. As such, these texts are intended for senior undergraduate and graduate students at the MS and PhD level; however, research scientists seeking an introduction to particular areas of physics will also benefit from the titles in this collection.

Springer
Berlin
Heidelberg
New York
Hong Kong
London
Milan
Paris
Tokyo

Toshiyuki Shiozawa

Classical Relativistic Electrodynamics

Theory of Light Emission and Application to Free Electron Lasers

With 73 Figures

Springer

Professor Toshiyuki Shiozawa
Osaka University
Department of Communication Engineering
Yamada-oka 2-1, Suita
Osaka, 565-0871, Japan
E-mail: shiozawa@comm.eng.osaka-u.ac.jp

Cataloging-in-Publication Data applied for

Bibliographic information published by Die Deutsche Bibliothek
Die Deutsche Bibliothek lists this publication in the Deutsche Nationalbibliografie; detailed bibliographic data is available in the Internet at <http://dnb.ddb.de>.

ISSN 1439-2674

ISBN 3-540-20623-X Springer-Verlag Berlin Heidelberg New York

Springer-Verlag is a part of Springer Science+Business Media

springeronline.com

Printed in Germany

Typesetting: Camera-ready by the author
Cover design: *design & production* GmbH, Heidelberg

Printed on acid-free paper SPIN 10954584 57/3141/ba 5 4 3 2 1 0

Preface

This book presents an advanced course of classical relativistic electrodynamics with application to the generation of high-power coherent radiation in the microwave to optical-wave regions. Specifically, it provides the reader with the basics of advanced electromagnetic theory and relativistic electrodynamics, guiding him step by step through the theory of free-electron lasers. In free-electron lasers, high-power coherent radiation is generated through the interaction of electrons, which are moving with relativistic velocity, with electromagnetic waves. As a prerequisite to this course, the reader should be familiar with the Maxwell equations and Newtonian mechanics. In addition, for the mathematical background, he should also have a good command of the three-dimensional vector calculus and the Fourier analysis. In the first preliminary chapter, the basics of advanced electromagnetic theory are reviewed. Then, in the second preliminary chapter, the foundations of relativistic electrodynamics are summarized within the framework of the special theory of relativity. The theoretical treatment in this chapter and throughout this book is fully developed by means of the usual three-dimensional vector calculus. Thus we could get a better insight into the physical meanings of the basic concepts in our theoretical development.

After the preliminary chapters for the basics of advanced electromagnetic theory and relativistic electrodynamics (Chaps. 1 and 2), Chap. 3 discusses electromagnetic radiation from a moving charged particle with emphasis on the synchrotron radiation and the Cherenkov radiation for the important subjects. In Chap. 4, we develop a macroscopic theory of relativistic electron beams. Specifically, we describe in detail the basic relations for small-signal fields propagated in a relativistic electron beam, namely, the basic field equations, constitutive relations, energy and momentum conservation relations and so forth. Then, on the basis of these basic relations, we investigate the properties of transverse and longitudinal waves which can be propagated in a relativistic electron beam. These two chapters (Chaps. 3 and 4) constitute the very basis for understanding the basic concepts of free-electron lasers.

The following chapters (Chaps. 5 ~ 7) are devoted to a detailed discussion of the basic amplification mechanisms for typical free-electron lasers, i.e., Cherenkov and wiggler-type free-electron lasers, with the aid of single-particle and collective approaches. Then, in the final chapter, the finite-difference time-domain (FDTD) method is presented for numerical analysis of beam-wave interaction, with many specific examples for the characteristics of the Cherenkov free-electron laser.

For the development of various concepts in this book, I heavily owe many authors through their published work. However, I have not tried to refer to all relevant publications. Instead, I listed only a limited number of representative books and journal articles at the end of the book, which are not exhaustive. Some of the illustrative numerical examples contained in this book have been taken from the research results obtained by myself and my coworkers. Some additional numerical examples have also been provided by Dr. Akimasa Hirata and Takahiko Adachi for this book. Including all these figures, most of the figures in this book have been prepared by Toshihiro Fujino with computer graphics. I would like to acknowledge their great efforts and collaborations.

Finally, this book can be recommended as a textbook or a reference book in the fields of advanced electromagnetic theory, relativistic electrodynamics, beam physics and plasma sciences for graduate students, or scientists and engineers in general.

Osaka, July 2003 Toshiyuki Shiozawa

Contents

1. Basic Electromagnetic Theory

1.1 Basic Field Equations in Vacuum

We first present the basic field equations for describing electromagnetic phenomena in an inertial frame of reference set up in vacuum. The basic physical quantities to represent electromagnetic fields in vacuum are the electric field intensity $\boldsymbol{E}$ [V/m], the magnetic-flux density $\boldsymbol{B}$ [T], and the source quantities to produce electromagnetic fields, namely, the current density $\boldsymbol{J}$ [A/m^2] and the charge density ρ [C/m^3]. These basic electromagnetic quantities are, in general, functions of position and time, and the relations between the field quantities, $\boldsymbol{E}$ and $\boldsymbol{B}$, and the source quantities, $\boldsymbol{J}$ and ρ are specified by the basic field equations, what is known as the Maxwell equations.

The integral form of the Maxwell equations in vacuum is given by the following expressions:

$$\oint_C \boldsymbol{E}\cdot d\boldsymbol{l} = -\frac{d}{dt}\int_S \boldsymbol{B}\cdot d\boldsymbol{s}, \tag{1.1.1}$$

$$\frac{1}{\mu_0}\oint_C \boldsymbol{B}\cdot d\boldsymbol{l} = \frac{d}{dt}\int_S \varepsilon_0\boldsymbol{E}\cdot d\boldsymbol{s} + \int_S \boldsymbol{J}\cdot d\boldsymbol{s}, \tag{1.1.2}$$

$$\oint_S \varepsilon_0\boldsymbol{E}\cdot d\boldsymbol{s} = \int_V \rho dv, \tag{1.1.3}$$

$$\oint_S \boldsymbol{B}\cdot d\boldsymbol{s} = 0, \tag{1.1.4}$$

where C in (1.1.1) and (1.1.2) denotes an arbitrary closed contour at rest in an inertial frame of reference, and S an arbitrary stationary open surface bounded by the closed contour C (see Fig. 1.1). In addition, S in (1.1.3) and (1.1.4) is an arbitrary closed surface at rest, and V the volume enclosed by the closed surface S (see Fig. 1.2). The constants ε_0 and μ_0 are the permittivity and permeability of vacuum, respectively.

In Fig. 1.1, $d\boldsymbol{l}$ represents a differential vector line element tangential to the contour C, and $d\boldsymbol{s}$ is a differential vector surface element normal to the surface S. In Fig. 1.2, on the other hand, $d\boldsymbol{s}$ denotes a differential vector surface element which is

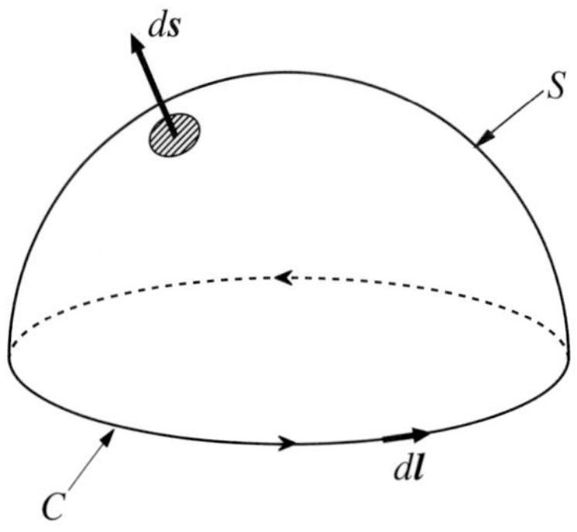

Fig. 1.1. A contour C and an open surface S bounded by C

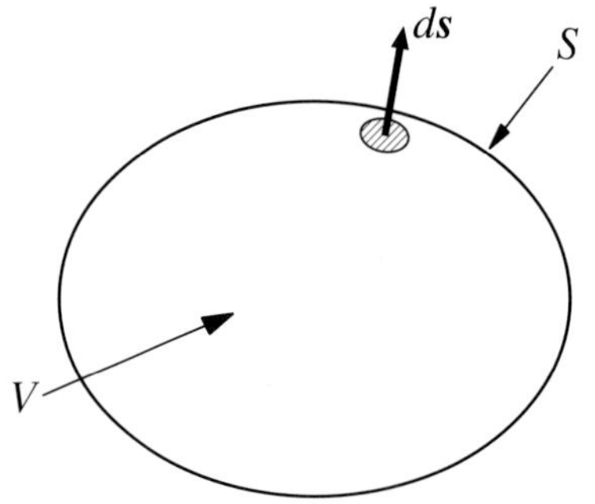

Fig. 1.2. A volume V enclosed by a closed surface S

outwardly normal to the closed surface S, and dv is a differential volume element. In Figs. 1.1 and 1.2, $\boldsymbol{ds}$ is also represented as $\boldsymbol{ds} = \boldsymbol{n}ds$, where $\boldsymbol{n}$ is a unit outward normal to the surface S and ds a differential surface element.

The first of the Maxwell equations, (1.1.1) expresses Faraday's induction law, which states that if the amount of the magnetic flux linking any closed contour C is temporally changed, the electromotive force is produced around C with the magnitude equal to the negative time rate of change of the magnetic flux linking the contour C. The second of the Maxwell equations, (1.1.2) is the extended version of Ampere's circuital law, what is called Ampere-Maxwell's law. According to this law, if current flows through any closed contour C or the amount of the electric flux linking the closed contour C is temporally changed, the magnetic field is produced around the closed contour C. In addition, the third and fourth of the Maxwell equations, (1.1.3) and (1.1.4) are Gauss' laws for the electric and magnetic fields, respectively. Gauss' law for the electric field states that the outward electric flux through any closed surface S is equal to the net amount of electric charge in the volume V enclosed by the closed surface S. On the other hand, Gauss' law for the magnetic field requires that the net outward magnetic flux through any closed surface S always vanish.

The current density $\boldsymbol{J}$ and the charge density ρ must satisfy the law of charge conservation given by

$$\oint_S \boldsymbol{J}\cdot d\boldsymbol{s} = -\frac{d}{dt}\int_V \rho dv, \tag{1.1.5}$$

where the left-hand side denotes the sum of the charge flowing out of the volume V enclosed by the closed surface S per second, while the right-hand side represents the time rate of decrease of the net charge contained in the volume V. Thus the equation (1.1.5) expresses the law of charge conservation. The law of charge conservation is one of the most fundamental laws of conservation in modern physics.

The law of charge conservation (1.1.5) can be deduced from the Maxwell equations. To verify this fact, let us divide a closed surface S enclosing a volume V into two open surfaces S_1 and S_2 bounded by a closed contour C on S, as shown in Fig. 1.3. Then, if we apply (1.1.2) to S_1 and S_2 bounded by C and add each of both sides of the resultant equations, we get

$$\frac{d}{dt}\oint_S \varepsilon_0 \boldsymbol{E}\cdot d\boldsymbol{s} + \oint_S \boldsymbol{J}\cdot d\boldsymbol{s} = 0. \tag{1.1.6}$$

Using the relation (1.1.3) in (1.1.6) leads to the law of charge conservation (1.1.5). Thus if we regard the Maxwell equations (1.1.1) ~ (1.1.4) as the independent basic equations, the law of charge conservation can be automatically derived from the Maxwell equations.

On the other hand, if we adopt the first two of the Maxwell equations, (1.1.1) and (1.1.2), together with the law of charge conservation (1.1.5), as the independent basic equations, then Gauss' law for the magnetic field (1.1.4) can be deduced from (1.1.1), and Gauss' law for the electric field (1.1.3) from (1.1.2) and (1.1.5). To prove these statements, let us first apply (1.1.1) to two open surfaces S_1 and S_2 bounded by a closed contour C on S as shown in Fig. 1.3, and add each of both sides of the resultant equations. Then we have

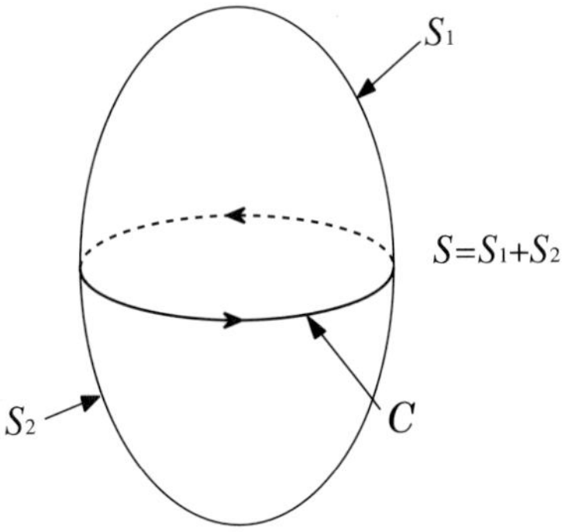

Fig. 1.3. Division of a closed surface S into two open surfaces S_1 and S_2 bounded by a closed contour C on S

$$\frac{d}{dt}\oint_S \boldsymbol{B}\cdot ds = 0. \tag{1.1.7}$$

Similarly, combining the relations (1.1.6) and (1.1.5), we obtain

$$\frac{d}{dt}\left[\oint_S \varepsilon_0 \boldsymbol{E}\cdot ds - \int_V \rho dv\right] = 0. \tag{1.1.8}$$

At a particular instant of time, we can always set $\boldsymbol{B} = 0$, $\boldsymbol{E} = 0$, and $\rho = 0$ on a closed surface S and in the volume V enclosed by the surface S. Integrating (1.1.7) and (1.1.8) with respect to time, together with these conditions, we immediately find Gauss' laws for the magnetic and electric fields (1.1.4) and (1.1.3).

Now, let us rewrite the Maxwell equations in integral form (1.1.1) ~ (1.1.4) into the corresponding differential form. For this purpose, we first apply Stokes' theorem to the left-hand sides of (1.1.1) and (1.1.2). Then we get

$$\int_S \left(\nabla\times\boldsymbol{E} + \frac{\partial \boldsymbol{B}}{\partial t}\right)\cdot ds = 0, \tag{1.1.9}$$

$$\int_S \left(\frac{1}{\mu_0}\nabla\times\boldsymbol{B} - \frac{\partial}{\partial t}\varepsilon_0\boldsymbol{E} - \boldsymbol{J}\right)\cdot ds = 0. \tag{1.1.10}$$

On the other hand, rewriting (1.1.3) and (1.1.4) by applying Gauss' theorem to the left-hand sides of them, we have

$$\int_V \left(\nabla\cdot\varepsilon_0\boldsymbol{E} - \rho\right)dv = 0, \tag{1.1.11}$$

$$\int_V \nabla\cdot\boldsymbol{B}dv = 0. \tag{1.1.12}$$

Equations (1.1.9) and (1.1.10) should hold for any open surface S bounded by an arbitrary closed contour C. Similarly, Eqs. (1.1.11) and (1.1.12) should apply for an arbitrary volume V. For these requirements to be satisfied, the inte-grands in (1.1.9) ~ (1.1.12) must vanish identically. Hence we find the following differential form of the Maxwell equations:

$$\nabla\times\boldsymbol{E} = -\frac{\partial\boldsymbol{B}}{\partial t}, \tag{1.1.13}$$

$$\nabla\times\frac{\boldsymbol{B}}{\mu_0} = \frac{\partial}{\partial t}\varepsilon_0\boldsymbol{E} + \boldsymbol{J}, \tag{1.1.14}$$

$$\nabla\cdot\varepsilon_0\boldsymbol{E} = \rho, \tag{1.1.15}$$

$$\nabla\cdot\boldsymbol{B} = 0. \tag{1.1.16}$$

In a similar fashion, the law of charge conservation in the integral form (1.1.5) can be recast into the corresponding differential form as

$$\nabla \cdot \boldsymbol{J} = -\frac{\partial \rho}{\partial t}. \tag{1.1.17}$$

Equation (1.1.17) is also referred to as the equation of continuity.

The integral forms of the Maxwell equations and the law of charge conservation, (1.1.1) ~ (1.1.4) and (1.1.5) can be applied to domains of space and time where electromagnetic quantities are continuous or discontinuous. On the other hand, it should be noted that the corresponding differential forms of the equations are applicable only to domains where electromagnetic quantities are differentiable as well as continuous.

1.2 Basic Field Equations in Material Media

It is well known that matter is composed of various kinds of particles such as atoms, molecules, electrons, or positive and negative ions. Hence, from the phenomenological point of view, matter is regarded as consisting of positive and negative charges. We find in matter two kinds of charges: one is the free charge which can move around rather freely in matter, and the other is the bound charge which is firmly bound to the atomic structure of matter and can move only slightly under the influence of external forces. The motion of these charges in matter produces electric currents. In particular, according to the classical theory of the atom, the electron bound to the nucleus undergoes an orbital motion around it, constituting an infinitesimal loop current. In addition to the orbital motion around the nucleus, the electron spins around its own axis, forming another infinitesimal loop current. These infinitesimal loop currents produced by the electron are equivalent to magnetic dipoles.

As is evident from the above discussion, we can regard matter as a collection of electric charges and currents distributed in vacuum, from the viewpoint of phenomenological electromagnetic theory. Thus the Maxwell equations for macroscopic fields in matter will result by incorporating the appropriately defined macroscopic charge and current densities distributed in vacuum into the Maxwell equations in vacuum presented in the foregoing section.

In the presence of polarized and magnetized matter, the macroscopic charge density ρ is represented as the sum of the free charge and polarization charge densities,

$$\rho = \rho_f - \nabla \cdot \boldsymbol{P}, \tag{1.2.1}$$

where ρ_f denotes the free charge density and $\boldsymbol{P}$ [C/m^2] the polarization which is defined by the average density of electric dipole moments.

On the other hand, the macroscopic current density is decomposed into the sum of the free current, the polarization current, and the magnetization current densities as

$$\boldsymbol{J} = \boldsymbol{J}_f + \frac{\partial \boldsymbol{P}}{\partial t} + \nabla \times \boldsymbol{M}, \tag{1.2.2}$$

$\boldsymbol{M}$ [A/m] representing the magnetization, which is the average density of magnetic dipole moments.

Inserting (1.2.1) and (1.2.2) for the charge and current densities in the Maxwell equations in vacuum, (1.1.1) ~ (1.1.4), and defining the new field vectors $\boldsymbol{D}$ (electric-flux density) [C/m^2] and $\boldsymbol{H}$ (magnetic field intensity) [A/m] to represent fields in polarized and magnetized matter as

$$\boldsymbol{D} = \varepsilon_0 \boldsymbol{E} + \boldsymbol{P}, \quad \boldsymbol{H} = \frac{\boldsymbol{B}}{\mu_0} - \boldsymbol{M}, \tag{1.2.3}$$

we obtain the Maxwell equations in polarized and magnetized matter.

The integral form of the Maxwell equations in polarized and magnetized matter corresponding to those in vacuum, (1.1.1) ~ (1.1.4) is given by

$$\oint_C \boldsymbol{E} \cdot d\boldsymbol{l} = -\frac{d}{dt}\int_S \boldsymbol{B} \cdot d\boldsymbol{s}, \tag{1.2.4}$$

$$\oint_C \boldsymbol{H} \cdot d\boldsymbol{l} = \frac{d}{dt}\int_S \boldsymbol{D} \cdot d\boldsymbol{s} + \int_S \boldsymbol{J}_f \cdot d\boldsymbol{s}, \tag{1.2.5}$$

$$\oint_S \boldsymbol{D} \cdot d\boldsymbol{s} = \int_V \rho_f dv, \tag{1.2.6}$$

$$\oint_S \boldsymbol{B} \cdot d\boldsymbol{s} = 0. \tag{1.2.7}$$

Similarly, for the differential form of the Maxwell equations in polarized and magnetized matter corresponding to (1.2.4) ~ (1.2.7), we find

$$\nabla \times \boldsymbol{E} = -\frac{\partial \boldsymbol{B}}{\partial t}, \tag{1.2.8}$$

$$\nabla \times \boldsymbol{H} = \frac{\partial \boldsymbol{D}}{\partial t} + \boldsymbol{J}_f, \tag{1.2.9}$$

$$\nabla \cdot \boldsymbol{D} = \rho_f, \tag{1.2.10}$$

$$\nabla \cdot \boldsymbol{B} = 0. \tag{1.2.11}$$

Equations (1.2.8) ~ (1.2.11) are the standard form of the Maxwell equations in material media.

1.3 Constitutive Relations

As described in the preceding section, only the two equations, (1.2.8) and (1.2.9) are independent among the Maxwell equations (1.2.8) ~ (1.2.11), while the other equations (1.2.10) and (1.2.11) can be deduced from the former two equations, together with the law of charge conservation (1.1.17). Therefore, if the current density $\boldsymbol{J}$ and the charge density ρ are specified, we can not uniquely determine the four field vectors $\boldsymbol{E}$, $\boldsymbol{B}$, $\boldsymbol{D}$, and $\boldsymbol{H}$ only from the Maxwell equations (1.2.8) ~ (1.2.11). In other words, for the Maxwell equations to have unique solutions for four field vectors, we need at least two more auxiliary equations to relate these field vectors, in addition to the Maxwell equations. These auxiliary equations are referred to as constitutive relations.

The constitutive relations are generally expressed as

$$\boldsymbol{D} = \boldsymbol{D}(\boldsymbol{E}, \boldsymbol{B}), \tag{1.3.1}$$

$$\boldsymbol{H} = \boldsymbol{H}(\boldsymbol{E}, \boldsymbol{B}), \tag{1.3.2}$$

of which the specific functional forms depend upon the properties of the space where electromagnetic phenomena occur, or upon those of the material media which occupy the space.

For the case where the current density $\boldsymbol{J}$ involves the component $\boldsymbol{J}_c$ which depends upon field vectors as is the case with the conduction current, the constitutive relation for $\boldsymbol{J}_c$ is represented in the form,

$$\boldsymbol{J}_c = \boldsymbol{J}_c(\boldsymbol{E}, \boldsymbol{B}). \tag{1.3.3}$$

The representations (1.3.1) ~ (1.3.3) mean that $\boldsymbol{D}$, $\boldsymbol{H}$, and $\boldsymbol{J}_c$ are functions of the field vectors $\boldsymbol{E}$ and $\boldsymbol{B}$. For simplicity, we treat here only linear electromagnetic systems where the constitutive relations (1.3.1) ~ (1.3.3) are linear functions of the field vectors $\boldsymbol{E}$ and $\boldsymbol{B}$.

First, the constitutive relations for vacuum or free space, which is the simplest of all, are given by

$$\boldsymbol{D} = \varepsilon_0 \boldsymbol{E}, \quad \boldsymbol{H} = \frac{1}{\mu_0}\boldsymbol{B}, \tag{1.3.4}$$

where ε_0 and μ_0 are the permittivity and permeability of vacuum, respectively, with the numerical values,

$$\varepsilon_0 = 8.854 \times 10^{-12} \text{ [F/m]}, \tag{1.3.5}$$

$$\mu_0 = 4\pi \times 10^{-7} \text{ [H/m]}. \tag{1.3.6}$$

For a linear, isotropic and nondispersive medium, the constitutive relations take the form,

$$\boldsymbol{D} = \varepsilon \boldsymbol{E}\,, \quad \boldsymbol{H} = \frac{1}{\mu}\boldsymbol{B}\,, \quad \boldsymbol{J}_c = \sigma \boldsymbol{E}\,, \tag{1.3.7}$$

ε, μ, and σ being the material constants referred to as the permittivity, the permeability, and the conductivity of the medium, respectively.

Those material media the material constants of which do not depend upon the magnitudes of field vectors are referred to as linear media, while those for which the material constants are independent of the directions of field vectors are called isotropic media. In addition, the relation (1.3.7) holds at each point in space, and may generally be functions of position. Those material media with uniform material constants are homogeneous media, while those the material constants of which vary spatially are inhomogeneous media.

For isotropic media, the material constants are scalar quantities as shown in (1.3.7). On the other hand, for the case where the material constants depend upon the directions of field vectors, they become tensorial quantities. Those material media of which the material constants depend upon the directions of field vectors are referred to as anisotropic media. For anisotropic media, the constitutive relations take the form,

$$\boldsymbol{D} = \hat{\varepsilon} \cdot \boldsymbol{E}\,, \quad \boldsymbol{H} = \hat{\mu}^{-1} \cdot \boldsymbol{B}\,, \quad \boldsymbol{J}_c = \hat{\sigma} \cdot \boldsymbol{E}\,, \tag{1.3.8}$$

where $\hat{\varepsilon}$, $\hat{\mu}$, and $\hat{\sigma}$ are the permittivity tensor, the permeability tensor, and the conductivity tensor, respectively, and $\hat{\mu}^{-1}$ denotes the inverse of $\hat{\mu}$.

Equations (1.3.8) are abbreviations for tensorial representations. For example, the first equation of (1.3.8) represents

$$\begin{bmatrix} D_x \\ D_y \\ D_z \end{bmatrix} = \begin{bmatrix} \varepsilon_{xx} & \varepsilon_{xy} & \varepsilon_{xz} \\ \varepsilon_{yx} & \varepsilon_{yy} & \varepsilon_{yz} \\ \varepsilon_{zx} & \varepsilon_{zy} & \varepsilon_{zz} \end{bmatrix} \begin{bmatrix} E_x \\ E_y \\ E_z \end{bmatrix}, \tag{1.3.9}$$

where D_i and E_i ($i = x, y, z$) denote the components of electric-flux density and electric field intensity in the rectangular coordinate system, and ε_{ij} ($i, j = x, y, z$) the elements of the permittivity tensor.

Finally, we touch briefly on another important nature of material media, namely, dispersion. For some material media, the polarization or magnetization at a particular instant of time depends not only upon the value of electric field intensity or magnetic field intensity at that instant of time, but also upon all their values prior to that instant of time. These media correspond to temporally dispersive media. In the frequency domain, the material constants of temporally dispersive media become frequency dependent as will be shown in a later section. The temporal dispersion is also referred to as the frequency dispersion. In addition to the temporal dispersion, we have another class of dispersion, namely, the spatial dispersion. In a spatially dispersive medium, the polarization and the magnetization at a particular position depend not only upon the values of electric and magnetic field intensities at that position but also upon all their values around that position. The spatial dispersion is also referred to as the wave number dispersion. For dispersive media, the constitutive relations in space and time domain cannot be expressed in such a simple form as in (1.3.7) or (1.3.8). Instead, they are represented by the superposition of all the contributions from the values of the polarization or magnetization prior to a particular instant of time and around a particular position. If we assume that the properties of a dispersive material medium are spatially and temporally invariant, the constitutive relations for the medium can be expressed as

$$\boldsymbol{D}(\boldsymbol{r},t) = \int \hat{\varepsilon}(\boldsymbol{r}-\boldsymbol{r}',t-t')\cdot \boldsymbol{E}(\boldsymbol{r}',t')d\boldsymbol{r}'dt', \tag{1.3.10}$$

$$\boldsymbol{H}(\boldsymbol{r},t) = \int \hat{\mu}^{-1}(\boldsymbol{r}-\boldsymbol{r}',t-t')\cdot \boldsymbol{B}(\boldsymbol{r}',t')d\boldsymbol{r}'dt', \tag{1.3.11}$$

in which $\hat{\varepsilon}(\boldsymbol{r}-\boldsymbol{r}',t-t')$ and $\hat{\mu}^{-1}(\boldsymbol{r}-\boldsymbol{r}',t-t')$ are functions of position and time inherent to a particular material medium, corresponding to the responses $\boldsymbol{D}(\boldsymbol{r},t)$ and $\boldsymbol{H}(\boldsymbol{r},t)$ to the impulsive fields $\boldsymbol{E}(\boldsymbol{r}',t')$ and $\boldsymbol{B}(\boldsymbol{r}',t')$ impressed at a particular position $\boldsymbol{r}'$ and at a particular time t'. Note that $\boldsymbol{r}$ and $\boldsymbol{r}'$ denote the position vectors for the field and source points, respectively.

Transforming the relations (1.3.10) and (1.3.11) in the space and time domains to the wave number and frequency domains by the Fourier analysis, we can express the constitutive relations for Fourier components of the field vectors in the same form as (1.3.7) or (1.3.8). Then, the material constants in the wave number and frequency domains become not only functions of wave number and frequency, but also complex quantities. The anisotropy and dispersion of material media become crucial in treating electromagnetic fields in such media as plasmas, electron beams, semiconductors, or ferrites. Later on in this book, we will discuss in detail the properties of relativistic electron beams which are a typical example of temporally and spatially dispersive media.

1.4 Boundary Conditions

As described in Sect. 1.1, the Maxwell equations in differential form can be applied only to domains of space and time where electromagnetic quantities are continuous and differentiable. Hence, in order to find solutions to the Maxwell equations in differential form in a domain including a surface at which field vectors suffer abrupt changes or discontinuities, we first solve them separately on each side of the surface where the field vectors are continuous. Then we interconnect the solutions obtained in the separate regions, by imposing the boundary conditions on them at the surface. The boundary conditions to be satisfied by field vectors at a surface where they change discontinuously can be found from the Maxwell equations in integral form (1.1.1) ~ (1.1.4), or (1.2.4) ~ (1.2.7).

At a surface S separating two regions I and II as illustrated in Fig. 1.4, let us derive the boundary conditions to be satisfied by field vectors from the Maxwell equations in integral form (1.2.4) ~ (1.2.7). For this purpose, we first consider a small thin cylindrical domain around a point P on the surface S, with top and bottom surfaces of area ΔS and height Δh between them. To the small

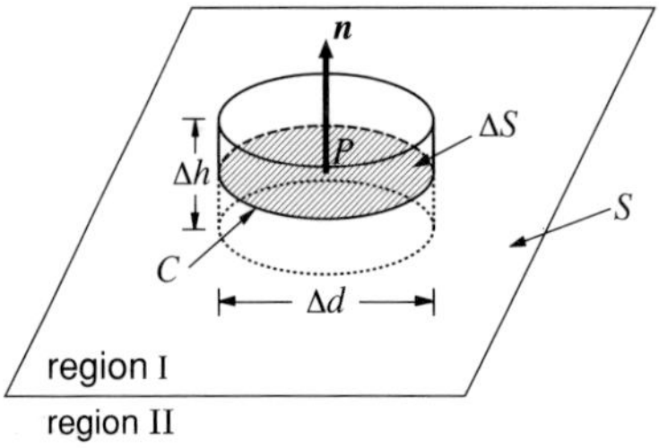

Fig. 1.4. A small cylindrical domain around a point P intersecting the boundary surface S

cylindrical domain, let us first apply the integral form of Gauss' law for the electric field, (1.2.6), and evaluate the limit when the small cylindrical domain shrinks around the point P, keeping the condition $\Delta h \ll \Delta d$, where Δd is the diameter of the top and bottom surfaces of the small cylindrical domain. Then, for the condition to be satisfied by the electric-flux density $\boldsymbol{D}$ at the point P on the boundary S, we get

$$\boldsymbol{n} \cdot (\boldsymbol{D}_1 - \boldsymbol{D}_2) = \xi, \tag{1.4.1}$$

where $\boldsymbol{D}_1$ and $\boldsymbol{D}_2$ are, respectively, the values of electric-flux density $\boldsymbol{D}$ at the point P on the sides of regions I and II, $\boldsymbol{n}$ a unit normal at the point P directed from region II to region I, and ξ the surface charge density, which is the amount of charge distributed over a unit area around the point P. Equation (1.4.1) shows that the difference between the normal components of electric-flux density should equal the surface charge density on the boundary.

Similarly, applying the integral form of Gauss' law for the magnetic field, (1.2.7) to the small domain illustrated in Fig. 1.4, and taking the same limiting process as for the preceding case, we find for the boundary condition for the normal components of magnetic-flux density $\boldsymbol{B}$,

$$\boldsymbol{n} \cdot (\boldsymbol{B}_1 - \boldsymbol{B}_2) = 0, \tag{1.4.2}$$

$\boldsymbol{B}_1$ and $\boldsymbol{B}_2$ being the values of magnetic-flux density $\boldsymbol{B}$ on the sides of regions I and II separated by the surface S. Equation (1.4.2) denotes the continuity of the normal components of magnetic-flux density on the boundary.

Next, consider a small rectangular contour C normal to the boundary surface S and intersecting S (see Fig. 1.5). The length of the sides of the rectangular contour normal to the boundary, Δh is assumed to be much less than the length of those parallel to the boundary, Δl. To the small rectangular contour C defined above and the open surface bounded by C, we apply the integral form of Faraday's induction law, (1.2.4) and evaluate the limit when the small rectangular contour C shrinks around the point P on the boundary surface S, keeping the condition $\Delta h \ll \Delta l$. Then, we obtain for the condition to be satisfied by the tangential component of electric field $\boldsymbol{E}$,

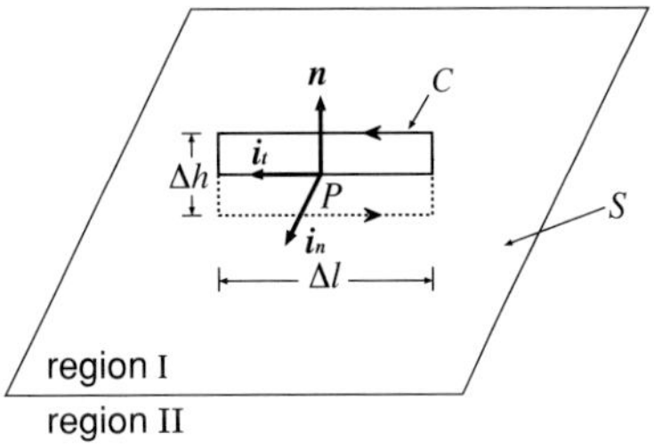

Fig. 1.5. A small rectangular contour around a point P intersecting the boundary surface S

$$\boldsymbol{i}_t \cdot (\boldsymbol{E}_1 - \boldsymbol{E}_2) = 0, \tag{1.4.3}$$

$\boldsymbol{E}_1$ and $\boldsymbol{E}_2$ denoting, respectively, the values of electric field $\boldsymbol{E}$ on the sides of regions I and II separated by the surface S, and $\boldsymbol{i}_t$ a unit vector parallel to the surface S.

The unit vector $\boldsymbol{i}_t$ can be expressed in terms of the unit vector $\boldsymbol{n}$, which is normal to the surface S and directed from region II to region I, and a unit vector $\boldsymbol{i}_n$, which is on the surface S and normal to both $\boldsymbol{n}$ and $\boldsymbol{i}_t$, as $\boldsymbol{i}_t = \boldsymbol{i}_n \times \boldsymbol{n}$. Hence, inserting this relation in (1.4.3) and rewriting the resultant equation with the aid of the identity for scalar triple products, we have

$$\boldsymbol{i}_n \cdot [\boldsymbol{n} \times (\boldsymbol{E}_1 - \boldsymbol{E}_2)] = 0. \tag{1.4.4}$$

The unit vector $\boldsymbol{i}_n$ can be directed in an arbitrary direction provided that it is on the surface S. Thus for the relation (1.4.4) to hold for an arbitrary $\boldsymbol{i}_n$, the following condition must be met on the surface S:

$$\boldsymbol{n} \times (\boldsymbol{E}_1 - \boldsymbol{E}_2) = 0. \tag{1.4.5}$$

Equation (1.4.5) means that the tangential components of electric field should be continuous across the boundary.

Finally, applying the integral form of Ampere-Maxwell's law, (1.2.5) to the small rectangular contour C and the open surface bounded by C defined in Fig. 1.5, we obtain the boundary condition for the tangential components of magnetic field $\boldsymbol{H}$,

$$\boldsymbol{n} \times (\boldsymbol{H}_1 - \boldsymbol{H}_2) = \boldsymbol{K}, \tag{1.4.6}$$

where $\boldsymbol{H}_1$ and $\boldsymbol{H}_2$ are, respectively, the values of magnetic field $\boldsymbol{H}$ on the sides of regions I and II separated by the surface S, and $\boldsymbol{K}$ the surface current density on the surface S, which is defined as the current flowing through a unit width on S. Equation (1.4.6) expresses that the difference between the tangential components of magnetic field on the boundary surface should be equal to the surface current density on the surface.

The boundary conditions to be satisfied by field vectors on the surface where they suffer discontinuities or abrupt changes are summarized as follows:

$$\boldsymbol{n} \times (\boldsymbol{E}_1 - \boldsymbol{E}_2) = 0, \tag{1.4.7}$$

$$\boldsymbol{n} \times (\boldsymbol{H}_1 - \boldsymbol{H}_2) = \boldsymbol{K}, \tag{1.4.8}$$

for the tangential components of electric and magnetic fields, and

$$\boldsymbol{n} \cdot (\boldsymbol{D}_1 - \boldsymbol{D}_2) = \xi, \tag{1.4.9}$$

$$\boldsymbol{n} \cdot (\boldsymbol{B}_1 - \boldsymbol{B}_2) = 0, \tag{1.4.10}$$

for the normal components of electric and magnetic flux densities. In (1.4.7) ~ (1.4.10), $\boldsymbol{n}$ is a unit vector normal to the boundary surface, which is directed from region II to region I, and ξ and $\boldsymbol{K}$ denote the surface charge and current densities.

For time-varying fields, the field vectors $\boldsymbol{E}$ and $\boldsymbol{B}$, or the field vectors $\boldsymbol{H}$ and $\boldsymbol{D}$ are interrelated to each other through the Maxwell equations. Hence the boundary condition for $\boldsymbol{B}$ is not independent of that for $\boldsymbol{E}$, while the boundary condition for $\boldsymbol{D}$ is not independent of that for $\boldsymbol{H}$. In other words, if the boundary condition for $\boldsymbol{E}$, (1.4.7) is satisfied, the condition (1.4.10) for $\boldsymbol{B}$ is automatically met from Faraday's induction law. Similarly, if the condition (1.4.8) for $\boldsymbol{H}$ is given, the condition (1.4.9) for $\boldsymbol{D}$ auto-matically holds from Ampere-Maxwell's law and the law of charge conservation.

For the special case where the region II is composed of a perfect conductor, we have the boundary conditions for field vectors,

$$\boldsymbol{n} \times \boldsymbol{E}_1 = 0, \tag{1.4.11}$$

$$\boldsymbol{n} \times \boldsymbol{H}_1 = \boldsymbol{K}. \tag{1.4.12}$$

For the purpose of determining electromagnetic fields outside a perfect conductor, it suffices to use the equation (1.4.11) only. Equation (1.4.12) is generally used to find the distribution of surface current on the surface of a perfect conductor.

1.5 Electromagnetic Potentials

The Maxwell equations are a set of coupled first-order differential equations that specifies the interrelation between electric and magnetic fields, and their source quantities, namely, charge and current densities. For some cases, this set of equations can be solved directly in terms of field vectors for given charge and current densities. For other cases, however, it is more convenient to solve the Maxwell equations with the aid of appropriately defined scalar and vector potentials. In this section, we define the scalar and vector potentials to represent electromagnetic fields.

From (1.1.16) or (1.2.11), we can first define the vector potential $\boldsymbol{A}$ [Wb/m] as

$$\boldsymbol{B} = \nabla \times \boldsymbol{A}. \tag{1.5.1}$$

Rewriting the first of the Maxwell equations, i.e., Faraday's induction law, (1.1.13) or (1.2.8) in terms of the vector potential $\boldsymbol{A}$, we get

$$\nabla \times (\boldsymbol{E} + \frac{\partial \boldsymbol{A}}{\partial t}) = 0, \tag{1.5.2}$$

which is equivalent to

$$\boldsymbol{E} + \frac{\partial \boldsymbol{A}}{\partial t} = -\nabla \phi, \tag{1.5.3}$$

or

$$\boldsymbol{E} = -\nabla \phi - \frac{\partial \boldsymbol{A}}{\partial t} \tag{1.5.4}$$

Equation (1.5.3) is the definition of the scalar potential ϕ [V] to represent the time-varying electric field.

The rest of the Maxwell equations, namely, Ampere-Maxwell's law (1.2.9) and Gauss' law for the electric field (1.2.10) can be rewritten in terms of the vector potential $\boldsymbol{A}$ defined by (1.5.1) and the scalar potential ϕ defined by (1.5.3) as

$$\nabla^2 \boldsymbol{A} - \varepsilon\mu \frac{\partial^2 \boldsymbol{A}}{\partial t^2} = -\mu \boldsymbol{J} + \nabla(\nabla \cdot \boldsymbol{A} + \varepsilon\mu \frac{\partial \phi}{\partial t}), \tag{1.5.5}$$

$$\nabla^2 \phi + \frac{\partial}{\partial t} \nabla \cdot \boldsymbol{A} = -\frac{\rho}{\varepsilon}, \tag{1.5.6}$$

where we used the constitutive relations for isotropic and nondispersive media, $\boldsymbol{D} = \varepsilon \boldsymbol{E}$ and $\boldsymbol{H} = (1/\mu)\boldsymbol{B}$ for simplicity.

As is well known, the vector and scalar potentials $\boldsymbol{A}$ and ϕ defined by (1.5.1) and (1.5.3) have some arbitrariness. Specifically, in place of a set of potentials ($\boldsymbol{A}$, ϕ), let us introduce another set of potentials $(\boldsymbol{A}', \phi')$ defined by

$$\boldsymbol{A}' = \boldsymbol{A} + \nabla \psi, \tag{1.5.7}$$

$$\phi' = \phi - \frac{\partial \psi}{\partial t}, \tag{1.5.8}$$

where ψ is an arbitrary continuous and differentiable scalar function. The transformation given by the above relations is referred to as a gauge transformation. The new set of potentials (1.5.7) and (1.5.8) evidently satisfies the two homogeneous Maxwell equations (1.2.8) and (1.2.11). There are infinite sets of vector and scalar potentials defined by the gauge transformation (1.5.7) and (1.5.8). Hence we can select those sets of potentials which meet the Lorentz condition,

$$\nabla \cdot \boldsymbol{A} + \varepsilon\mu \frac{\partial \phi}{\partial t} = 0. \tag{1.5.9}$$

Then, from (1.5.5) and (1.5.6), we obtain for the wave equations for $\boldsymbol{A}$ and ϕ,

$$\nabla^2 \boldsymbol{A} - \frac{1}{u^2} \frac{\partial^2 \boldsymbol{A}}{\partial t^2} = -\mu \boldsymbol{J}, \tag{1.5.10}$$

$$\nabla^2 \phi - \frac{1}{u^2}\frac{\partial^2 \phi}{\partial t^2} = -\frac{\rho}{\varepsilon}, \tag{1.5.11}$$

with

$$u = \frac{1}{\sqrt{\varepsilon \mu}}, \tag{1.5.12}$$

u being the speed of light in an unbounded medium with material constants ε and μ. For vacuum, in particular, $\varepsilon = \varepsilon_0$, $\mu = \mu_0$, and $u = c$, where c is the speed of light in vacuum the numerical value of which is given by

$$c = \frac{1}{\sqrt{\varepsilon_0 \mu_0}} = 2.99792458 \times 10^8 \text{ [m/s]}. \tag{1.5.13}$$

1.6 Field Equations in the Frequency Domain

According to the Fourier analysis, electromagnetic fields temporally varying in an arbitrary manner can be analyzed as a superposition of elementary fields varying sinusoidally in time at a single frequency. Thus the electromagnetic field temporally varying at a single frequency is of great importance.

For the analysis of the electromagnetic field varying sinusoidally in time at a single frequency, it is convenient to use the complex representation. For example, a real electric field vector $\boldsymbol{E}$ varying sinusoidally in time with angular frequency ω is expressed as

$$\boldsymbol{E}(\boldsymbol{r},t) = \mathrm{Re}[\tilde{\boldsymbol{E}}(\boldsymbol{r},\omega)e^{j\omega t}] = \frac{1}{2}[\tilde{\boldsymbol{E}}(\boldsymbol{r},\omega)e^{j\omega t} + \tilde{\boldsymbol{E}}^*(\boldsymbol{r},\omega)e^{-j\omega t}], \tag{1.6.1}$$

where $\tilde{\boldsymbol{E}}(\boldsymbol{r},\omega)$ is a function of position and frequency, $\tilde{\boldsymbol{E}}^*(\boldsymbol{r},\omega)$ is the complex conjugate of $\tilde{\boldsymbol{E}}(\boldsymbol{r},\omega)$, and Re denotes the operation of taking the real part of the following function.

The other field vectors $\boldsymbol{B}$, $\boldsymbol{D}$, and $\boldsymbol{H}$, and the current and charge densities $\boldsymbol{J}$ and ρ can also be represented in the same complex form as (1.6.1). Then, inserting these complex representations for field vectors and source quantities in (1.2.8) ~ (1.2.11), we get the following complex representation for the Maxwell equations in material media:

$$\nabla \times \tilde{\boldsymbol{E}} = -j\omega \tilde{\boldsymbol{B}}, \tag{1.6.2}$$

$$\nabla \times \tilde{\boldsymbol{H}} = j\omega \tilde{\boldsymbol{D}} + \tilde{\boldsymbol{J}}, \tag{1.6.3}$$

$$\nabla \cdot \tilde{\boldsymbol{D}} = \tilde{\rho}, \tag{1.6.4}$$

$$\nabla \cdot \tilde{\boldsymbol{B}} = 0. \tag{1.6.5}$$

Similarly, we have for the complex representation for the law of charge conservation,

$$\nabla \cdot \tilde{\boldsymbol{J}} = -j\omega\tilde{\rho}. \tag{1.6.6}$$

Next, let us consider the constitutive relations in the frequency domain. For nondispersive media, the constitutive relations (1.3.7) and (1.3.8) can be used also in the frequency domain. However, all the material constants appearing in the constitutive relations in the time domain should be real quantities, while those in the frequency domain are allowed to be complex quantities. Hence, in the frequency domain, we can define the equivalent permittivity ε' and the equivalent permittivity tensor $\hat{\varepsilon}'$ as

$$\varepsilon' = \varepsilon - j\frac{\sigma}{\omega}, \tag{1.6.7}$$

$$\hat{\varepsilon}' = \hat{\varepsilon} - j\frac{\hat{\sigma}}{\omega}. \tag{1.6.8}$$

For dispersive media, on the other hand, the constitutive relations in the frequency domain are given by

$$\tilde{\boldsymbol{D}} = \hat{\varepsilon}(\omega)\cdot\tilde{\boldsymbol{E}}, \quad \tilde{\boldsymbol{H}} = \hat{\mu}^{-1}(\omega)\cdot\tilde{\boldsymbol{B}}, \tag{1.6.9}$$

where we have taken into account only the temporal dispersion. The complex counterparts of the permittivity tensor and the inverse of the permeability tensor, $\hat{\varepsilon}(\omega)$ and $\hat{\mu}^{-1}(\omega)$ are expressed as

$$\hat{\varepsilon}(\omega) = \int_0^\infty \hat{\varepsilon}(\tau)e^{-j\omega\tau}d\tau, \tag{1.6.10}$$

$$\hat{\mu}^{-1}(\omega) = \int_0^\infty \hat{\mu}^{-1}(\tau)e^{-j\omega\tau}d\tau, \tag{1.6.11}$$

which are obtained by inserting the complex representation (1.6.1) for the electric field vector $\boldsymbol{E}$ and similar expressions for the other field vectors in (1.3.10) and (1.3.11). Equations (1.6.10) and (1.6.11) represent the Fourier transforms for the temporal functions $\hat{\varepsilon}(\tau)$ and $\hat{\mu}^{-1}(\tau)$. It should be noted that $\hat{\varepsilon}(\tau)$ and $\hat{\mu}^{-1}(\tau)$ are real quantities while $\hat{\varepsilon}(\omega)$ and $\hat{\mu}^{-1}(\omega)$ are generally complex.

The field equations (1.6.2) ~ (1.6.5) and the constitutive relations (1.6.9) have been derived for electromagnetic fields varying sinusoidally in time with a single frequency ω. For electromagnetic fields varying temporally in an arbitrary manner, we decompose them into a sum of elementary fields varying sinusoidally with single frequencies, according to the Fourier analysis. Then, each elementary field satisfies the same field equations and constitutive relations as given above for a single frequency field.

Specifically, for fields varying arbitrarily in time, the electric field vector can be expressed in terms of the Fourier integral instead of (1.6.1) as

$$\boldsymbol{E}(\boldsymbol{r},t) = \frac{1}{2\pi}\int_{-\infty}^{\infty} \tilde{\boldsymbol{E}}(\boldsymbol{r},\omega) e^{j\omega t} d\omega . \tag{1.6.12}$$

The other field vectors, and the charge and current densities can also be represented in a similar fashion in terms of the Fourier integrals. Then, introducing these expressions for field vectors, and for charge and current densities in terms of the Fourier integrals in the Maxwell equations and the constitutive relations in time domain, we get the same field equations and constitutive relations as (1.6.2) ~ (1.6.5) and (1.6.9). Thus, Eqs. (1.6.2) ~ (1.6.5) and (1.6.9) prove to be the basic equations for the analysis of electromagnetic fields in frequency domain.

1.7 Energy Conservation Relations

In this section, let us discuss the law of energy conservation for electromagnetic fields in material media. First, we consider energy conservation relations for electromagnetic fields in nondispersive media. Then, we treat a more general case for electromagnetic fields in dispersive media.

1.7.1 Energy Conservation Law in Nondispersive Media

Subtracting the scalar product of the electric field $\boldsymbol{E}$ with (1.2.9) from the scalar product of the magnetic field $\boldsymbol{H}$ with (1.2.8), we obtain

$$\boldsymbol{H}\cdot\nabla\times\boldsymbol{E} - \boldsymbol{E}\cdot\nabla\times\boldsymbol{H} = -\boldsymbol{H}\cdot\frac{\partial \boldsymbol{B}}{\partial t} - \boldsymbol{E}\cdot\frac{\partial \boldsymbol{D}}{\partial t} - \boldsymbol{E}\cdot\boldsymbol{J}. \tag{1.7.1}$$

Rewriting the left-hand side of (1.7.1) with the aid of the vector identity, we get

$$\nabla\cdot(\boldsymbol{E}\times\boldsymbol{H}) = -\boldsymbol{H}\cdot\frac{\partial \boldsymbol{B}}{\partial t} - \boldsymbol{E}\cdot\frac{\partial \boldsymbol{D}}{\partial t} - \boldsymbol{E}\cdot\boldsymbol{J}. \tag{1.7.2}$$

For nondispersive media, the constitutive relations are given by (1.3.7) or (1.3.8). In addition, we decompose the current density $\boldsymbol{J}$ into two components, one for the externally impressed current, $\boldsymbol{J}_s$ and the other for the conduction current, $\boldsymbol{J}_c$ which is dependent on electromagnetic fields in material media. Then, Eq. (1.7.2) can be rewritten in the form

$$\nabla\cdot\boldsymbol{S} + \frac{\partial w}{\partial t} + p_d = -\boldsymbol{E}\cdot\boldsymbol{J}_s, \tag{1.7.3}$$

where

$$\boldsymbol{S} = \boldsymbol{E}\times\boldsymbol{H}, \tag{1.7.4}$$

$$w = w_e + w_m, \tag{1.7.5}$$

$$w_e = \frac{1}{2}\boldsymbol{E}\cdot\boldsymbol{D},\quad w_m = \frac{1}{2}\boldsymbol{H}\cdot\boldsymbol{B}, \tag{1.7.6}$$

$$p_d = \boldsymbol{E} \cdot \boldsymbol{J}_c. \tag{1.7.7}$$

The vector $\boldsymbol{S}$ [W/m^2] is referred to as the Poynting vector which represents the power flow density, namely, the power flow passing through unit area perpendicular to the vector $\boldsymbol{S}$. In addition, the scalar w [J/m^3] denotes the electromagnetic energy density which corresponds to the electromagnetic energy stored per unit volume. It is the sum of the electric energy density w_e and the magnetic energy density w_m. Furthermore, the scalar p_d [W/m^3] expresses the power per unit volume transferred from the electromagnetic field to the Joule heat, namely, the density of consumed power. On the other hand, the right-hand side of (1.7.3) is equal to the energy per unit volume supplied from the external source per unit time, namely, the source power density.

For the special case of isotropic and nondispersive media, w_e, w_m, and Q are expressed with the aid of (1.3.7) as

$$w_e = \frac{1}{2}\varepsilon|\boldsymbol{E}|^2, \quad w_m = \frac{1}{2}\mu|\boldsymbol{H}|^2, \quad p_d = \sigma|\boldsymbol{E}|^2. \tag{1.7.8}$$

For anisotropic and nondispersive media, on the other hand, they take the form,

$$w_e = \frac{1}{2}\boldsymbol{E}\cdot\hat{\varepsilon}\cdot\boldsymbol{E}, \quad w_m = \frac{1}{2}\boldsymbol{H}\cdot\hat{\mu}\cdot\boldsymbol{H}, \quad p_d = \boldsymbol{E}\cdot\hat{\sigma}\cdot\boldsymbol{E}, \tag{1.7.9}$$

in which $\boldsymbol{E}\cdot\hat{\varepsilon}\cdot\boldsymbol{E}$, for example, denotes the scalar product of $\boldsymbol{E}$ and $\boldsymbol{D}$, namely, the scalar product of $\boldsymbol{E}$ and $\hat{\varepsilon}\cdot\boldsymbol{E}$.

The differential form of the law of energy conservation in nondispersive media (1.7.3) can be rewritten in the corresponding integral form. For this purpose, we integrate (1.7.3) in the volume V enclosed by a closed surface S as shown in Fig. 1.2, and transform the first term on the left-hand side of the resultant equation to the equivalent surface integral with the aid of Gauss' theorem. Then, we obtain the following integral form of the energy conservation law in nondispersive media:

$$\oint_S \boldsymbol{S}\cdot d\boldsymbol{s} + \frac{d}{dt}\int_V w\,dv + \int_V p_d\,dv = -\int_V \boldsymbol{E}\cdot\boldsymbol{J}_s\,dv. \tag{1.7.10}$$

The first term on the left-hand side of (1.7.10) expresses the energy flowing out of the volume V through the closed surface S per unit time. In addition, the second term represents the temporal increment of the electromagnetic energy stored in the volume V. Furthermore, the third term corresponds to the amount of energy transferred from the electromagnetic energy to the Joule heat per unit time in the volume V. On the other hand, the right-hand side of (1.7.10) denotes the energy supplied per unit time from an external source in the volume V. Thus, the equation (1.7.10) means that the energy equal to the sum of the energy lost from the volume V per unit time and the temporal increment of energy stored in the volume is supplied from an external source in the volume per unit time. In other words, it represents the law of energy conservation for the electromagnetic field. The equations (1.7.3) and (1.7.10) are referred to as Poynting's theorem.

Let us now assume that we have no electromagnetic fields present in the volume V in the limit $t \to -\infty$. Then, integrating the second term on the left-hand side of (1.7.10) from $t = -\infty$ to $t = t$, we get for the energy stored in the volume V by the time t, W,

$$W = \int_V w dv = W_e + W_m , \tag{1.7.11}$$

$$W_e = \int_V w_e dv, \tag{1.7.12}$$

$$W_m = \int_V w_m dv, \tag{1.7.13}$$

where W_e denotes the total energy stored in the electric field, and W_m the total energy stored in the magnetic field.

1.7.2 Energy Conservation Law in Dispersive Media

Let us consider the law of energy conservation for electromagnetic fields in temporally dispersive media. In temporally dispersive media, we cannot use the relation (1.7.8) or (1.7.9) for the electromagnetic energy densities. This is because in temporally dispersive media, as described in Sect. 1.3, the polarization or magneti-zation at a particular instant of time depends not only upon the value of electric field intensity or magnetic field intensity at that instant of time, but also upon all their values prior to that instant of time. In other words, the constitutive relations for temporally dispersive media in time domain cannot be expressed as (1.3.7) or (1.3.8) for nondispersive media. This is the reason why the representation (1.7.8) or (1.7.9) for the electromagnetic energy densities cannot be used for dispersive media.

To formulate correctly the law of energy conservation for electromagnetic fields in temporally dispersive media, let us treat a quasi-sinusoidally oscillating electromagnetic field with slowly varying amplitudes instead of a purely sinusoidal field oscillating with a single frequency. The quasi-sinusoidal fields defined above include, as a special case, a purely sinusoidal field varying temporally with a single frequency.

For a quasi-sinusoidal field with slowly varying amplitudes, we formulate the law of energy conservation which dictates the relation among the time-average power flow density, the time-average energy density, and the time-average power loss density. For this purpose, let us calculate the time-average of (1.7.2) over one period of time for a quasi-sinusoidal field with slowly varying amplitudes,

$$\nabla \cdot \langle \boldsymbol{E} \times \boldsymbol{H} \rangle = -\left\langle \boldsymbol{E} \cdot \frac{\partial \boldsymbol{D}}{\partial t} \right\rangle - \left\langle \boldsymbol{H} \cdot \frac{\partial \boldsymbol{B}}{\partial t} \right\rangle - \langle \boldsymbol{E} \cdot \boldsymbol{J}_s \rangle, \tag{1.7.14}$$

where the angle brackets denote taking the time-average of the function enclosed in them, and $\boldsymbol{J}_s$ denotes an externally applied source current.

To calculate the time-average of product quantities in (1.7.14), we note the following formula for the time-average of the product of two physical quantities f and g:

$$\langle fg \rangle = \frac{1}{2}\mathrm{Re}(\tilde{g}^* \tilde{f}) = \frac{1}{2}\mathrm{Re}(\tilde{f}^* \tilde{g}), \tag{1.7.15}$$

which can be readily derived by expressing f and g in complex form as (1.6.1).

Now, let us make some preparations for the calculation of the time-average of the product quantities in (1.7.14). First, the electric field vector $\boldsymbol{E}$ for a quasi-sinusoidal field with slowly varying amplitudes can be represented in complex form as

$$\boldsymbol{E}(\boldsymbol{r},t) = \mathrm{Re}[\tilde{\boldsymbol{E}}(\boldsymbol{r},t)e^{j\omega t}], \tag{1.7.16}$$

where $\tilde{\boldsymbol{E}}(\boldsymbol{r},t)$ denotes a complex amplitude varying slowly in time. Rewriting the right-hand side of (1.7.16) in terms of the Fourier transform, we obtain

$$\tilde{\boldsymbol{E}}(\boldsymbol{r},t)e^{j\omega t} = \frac{1}{2\pi}\int_{-\infty}^{\infty} \overline{\boldsymbol{E}}(\boldsymbol{r},\omega')e^{j\omega' t}d\omega'. \tag{1.7.17}$$

Note that the Fourier component $\overline{\boldsymbol{E}}(\boldsymbol{r},\omega')$ does not vanish only around $\omega' = \omega$.

Expressing the electric-flux density $\boldsymbol{D}$ in complex form in the same manner as for $\boldsymbol{E}$ in (1.7.17), and using the constitutive relation $\overline{\boldsymbol{D}}(\boldsymbol{r},\omega') = \hat{\varepsilon}(\omega')\cdot\overline{\boldsymbol{E}}(\boldsymbol{r},\omega')$ for each Fourier component, we find for the electric-flux density,

$$\tilde{\boldsymbol{D}}(\boldsymbol{r},t)e^{j\omega t} = \frac{1}{2\pi}\int_{-\infty}^{\infty} \hat{\varepsilon}(\omega')\cdot\overline{\boldsymbol{E}}(\boldsymbol{r},\omega')e^{j\omega' t}d\omega'. \tag{1.7.18}$$

Expanding $\hat{\varepsilon}(\omega')$ in Taylor series around $\omega' = \omega$, and leaving the terms to the first order in $(\omega' - \omega)$, we get

$$\hat{\varepsilon}(\omega') = \hat{\varepsilon}(\omega) + \frac{\partial\hat{\varepsilon}(\omega)}{\partial\omega}(\omega' - \omega). \tag{1.7.19}$$

Inserting (1.7.19) in (1.7.18) yields

$$\tilde{\boldsymbol{D}}(\boldsymbol{r},t) = \hat{\varepsilon}(\omega)\cdot\tilde{\boldsymbol{E}}(\boldsymbol{r},t) - j\frac{\partial\hat{\varepsilon}(\omega)}{\partial\omega}\cdot\frac{\partial\tilde{\boldsymbol{E}}(\boldsymbol{r},t)}{\partial t}. \tag{1.7.20}$$

Assuming that $\tilde{\boldsymbol{E}}(\boldsymbol{r},t)$ varies slowly in time, we can express the time derivative of $\boldsymbol{D}(\boldsymbol{r},t)$ with the aid of (1.7.20) as

$$\frac{\partial\boldsymbol{D}}{\partial t} = \mathrm{Re}\left[\left(j\omega\tilde{\boldsymbol{D}} + \frac{\partial\tilde{\boldsymbol{D}}}{\partial t}\right)e^{j\omega t}\right] = \mathrm{Re}\left[\left(j\omega\hat{\varepsilon}\cdot\tilde{\boldsymbol{E}} + \frac{\partial\omega\hat{\varepsilon}}{\partial\omega}\cdot\frac{\partial\tilde{\boldsymbol{E}}}{\partial t}\right)e^{j\omega t}\right]. \tag{1.7.21}$$

We are now prepared for calculating the first term on the right-hand side of (1.7.14). Using (1.7.16) and (1.7.21) in the formula (1.7.15), we get

$$\left\langle \boldsymbol{E} \cdot \frac{\partial \boldsymbol{D}}{\partial t} \right\rangle = \frac{1}{2} \mathrm{Re} \left[\tilde{\boldsymbol{E}} \cdot \left(-j\omega \hat{\varepsilon}^* \cdot \tilde{\boldsymbol{E}}^* + \frac{\partial \omega \hat{\varepsilon}^*}{\partial \omega} \cdot \frac{\partial \tilde{\boldsymbol{E}}^*}{\partial t} \right) \right]. \tag{1.7.22}$$

To rewrite the right-hand side of (1.7.22), let us decompose the elements ε_{ij} ($i, j = x, y, z$) of the permittivity tensor $\hat{\varepsilon}$ into Hermitian parts $\varepsilon_{ij}{}^h$ associated with stored energy and anti-Hermitian parts $\varepsilon_{ij}{}^a$ associated with energy dissipation.

$$\varepsilon_{ij} = \varepsilon_{ij}{}^h + \varepsilon_{ij}{}^a, \tag{1.7.23}$$

$$\varepsilon_{ij}{}^h = \frac{1}{2}\left(\varepsilon_{ij} + \varepsilon_{ji}{}^*\right), \tag{1.7.24}$$

$$\varepsilon_{ij}{}^a = \frac{1}{2}\left(\varepsilon_{ij} - \varepsilon_{ji}{}^*\right), \tag{1.7.25}$$

where $\varepsilon_{ij}{}^h$ and $\varepsilon_{ij}{}^a$ are assumed to satisfy the condition $\left|\varepsilon_{ij}{}^h\right| >> \left|\varepsilon_{ij}{}^a\right|$.

With the aid of (1.7.23), Eq. (1.7.22) is rewritten in the form,

$$\left\langle \boldsymbol{E} \cdot \frac{\partial \boldsymbol{D}}{\partial t} \right\rangle = \frac{\partial}{\partial t} \left[\frac{1}{4} \tilde{\boldsymbol{E}} \cdot \frac{\partial \omega (\hat{\varepsilon}^h)^*}{\partial \omega} \cdot \tilde{\boldsymbol{E}}^* \right] + \mathrm{Im} \left[\frac{1}{2} \omega \tilde{\boldsymbol{E}} \cdot (\hat{\varepsilon}^a)^* \cdot \tilde{\boldsymbol{E}}^* \right], \tag{1.7.26}$$

$\hat{\varepsilon}^h$ and $\hat{\varepsilon}^a$ denoting, respectively, the tensors with $\varepsilon_{ij}{}^h$ and $\varepsilon_{ij}{}^a$ as their elements, and Im representing taking the imaginary part of the following function.

It should be noted that Eq. (1.7.26) holds good only if the condition $\left|\varepsilon_{ij}{}^h\right| >> \left|\varepsilon_{ij}{}^a\right|$ is satisfied. If the similar condition $\left|\mu_{ij}{}^h\right| >> \left|\mu_{ij}{}^a\right|$ for the permeability tensor is satisfied, we find an expression similar to (1.7.26) for the time-average of the second term on the right-hand side of (1.7.14). Thus, for a quasi-sinusoidal field with time varying amplitudes propagated in temporally dispersive media with vanishingly small loss, we get, from (1.7.14), the following energy conservation relation:

$$\nabla \cdot \langle \boldsymbol{S} \rangle + \frac{\partial \langle w \rangle}{\partial t} + \langle p_d \rangle = -\langle q \rangle, \tag{1.7.27}$$

with

$$\langle \boldsymbol{S} \rangle = \frac{1}{2} \mathrm{Re}[\tilde{\boldsymbol{E}} \times \tilde{\boldsymbol{H}}^*], \tag{1.7.28}$$

$$\langle w \rangle = \langle w_e \rangle + \langle w_m \rangle, \tag{1.7.29}$$

$$\langle w_e \rangle = \frac{1}{4} \tilde{\boldsymbol{E}} \cdot \frac{\partial \omega (\hat{\varepsilon}^h)^*}{\partial \omega} \cdot \tilde{\boldsymbol{E}}^*, \quad \langle w_m \rangle = \frac{1}{4} \tilde{\boldsymbol{H}} \cdot \frac{\partial \omega (\hat{\mu}^h)^*}{\partial \omega} \cdot \tilde{\boldsymbol{H}}^*, \tag{1.7.30}$$

$$\langle p_d \rangle = \frac{1}{2} \omega \, \mathrm{Im} \left[\tilde{\boldsymbol{E}} \cdot (\hat{\varepsilon}^a)^* \cdot \tilde{\boldsymbol{E}}^* + \tilde{\boldsymbol{H}} \cdot (\hat{\mu}^a)^* \cdot \tilde{\boldsymbol{H}}^* \right], \tag{1.7.31}$$

$$\langle q\rangle = \frac{1}{2}\mathrm{Re}\left(\tilde{\boldsymbol{E}}\cdot\tilde{\boldsymbol{J}}_s^*\right). \tag{1.7.32}$$

In (1.7.27), $<w_e>$ and $<w_m>$ denote, respectively, the time-average densities of electric and magnetic energies stored in temporally dispersive media, and $<p_d>$ is the time-average density of power absorbed in temporally dispersive media. From the expression for $<p_d>$, we find for the condition for material media to be loss free,

$$\varepsilon_{ij}{}^a = \mu_{ij}{}^a = 0. \tag{1.7.33}$$

From the definition in (1.7.25), the condition (1.7.33) means

$$\varepsilon_{ij} = \varepsilon_{ji}{}^*, \quad \mu_{ij} = \mu_{ji}{}^*. \tag{1.7.34}$$

For isotropic media, in particular, the above condition represents that the permittivity and permeability of material media are real quantities.

Finally, it should be emphasized again that the energy conservation relation given by (1.7.27) with (1.7.28) ~ (1.7.32) are valid only for temporally dispersive media with vanishingly small loss. In addition, considering the limiting case where the amplitudes of a quasi-monochromatic field vary sufficiently slowly, we find that the energy conservation relation (1.7.27) with (1.7.28) ~ (1.7.32) can be applied also to a purely monochromatic field. Furthermore, in the absencc of dispersion, the relation (1.7.27) with (1.7.28) ~ (1.7.32) corresponds to the time-averages of (1.7.3) with (1.7.4) ~ (1.7.7) for nondispersive media.

1.8 Plane Waves

From the Maxwell equations (1.2.8) ~ (1.2.11) in material media with the charge and current densities set equal to zero, together with the constitutive relations (1.3.7), we can get the following homogeneous wave equations for electric and magnetic fields in an unbounded medium with the permittivity ε and the permeability μ:

$$\nabla^2\boldsymbol{E} - \frac{1}{u^2}\frac{\partial^2\boldsymbol{E}}{\partial t^2} = 0, \tag{1.8.1}$$

$$\nabla^2\boldsymbol{H} - \frac{1}{u^2}\frac{\partial^2\boldsymbol{H}}{\partial t^2} = 0, \tag{1.8.2}$$

where u is the speed of light in an unbounded medium with the permittivity ε and the permeability μ defined in (1.5.12).

To solve (1.8.1) and (1.8.2), let us express the electric and magnetic field vectors $\boldsymbol{E}$ and $\boldsymbol{H}$ in complex form as in (1.6.1),

$$\boldsymbol{E}(\boldsymbol{r},t) = \mathrm{Re}[\tilde{\boldsymbol{E}}(\boldsymbol{r},\omega)e^{j\omega t}], \tag{1.8.3}$$

$$\boldsymbol{H}(\boldsymbol{r},t) = \mathrm{Re}[\tilde{\boldsymbol{H}}(\boldsymbol{r},\omega)e^{j\omega t}]. \tag{1.8.4}$$

Then, inserting these complex representations for $\boldsymbol{E}$ and $\boldsymbol{H}$ in the wave equations (1.8.1) and (1.8.2), we obtain the following homogeneous Helmholtz equations:

$$\nabla^2 \tilde{\boldsymbol{E}} + \frac{\omega^2}{u^2}\tilde{\boldsymbol{E}} = 0, \tag{1.8.5}$$

$$\nabla^2 \tilde{\boldsymbol{H}} + \frac{\omega^2}{u^2}\tilde{\boldsymbol{H}} = 0. \tag{1.8.6}$$

We seek plane wave solutions to the homogeneous Helmholtz equations (1.8.5) and (1.8.6). The plane wave is defined as a transverse electromagnetic wave for which equiphase surfaces or wave fronts are infinite planes perpendicular to the direction of wave propagation. In addition, the amplitudes of electric and magnetic field vectors for the plane wave are constant on the equiphase surface. Hence, for plane waves, we can assume solutions to (1.8.5) and (1.8.6) in the form,

$$\tilde{\boldsymbol{E}}(\boldsymbol{r},\omega) = \tilde{\boldsymbol{E}}_0 e^{-j\boldsymbol{k}\cdot\boldsymbol{r}}, \tag{1.8.7}$$

$$\tilde{\boldsymbol{H}}(\boldsymbol{r},\omega) = \tilde{\boldsymbol{H}}_0 e^{-j\boldsymbol{k}\cdot\boldsymbol{r}}, \tag{1.8.8}$$

where $\tilde{\boldsymbol{E}}_0$ and $\tilde{\boldsymbol{H}}_0$ are constant vectors, and $\boldsymbol{k}$ is also a constant vector directed in the direction of wave propagation. By substitution of (1.8.7) and (1.8.8) into (1.8.5) and (1.8.6), the magnitude of $\boldsymbol{k}$ proves to be equal to ω/u.

Next, from the Maxwell equations (1.6.2) ~ (1.6.5), the relations among $\tilde{\boldsymbol{E}}_0$, $\tilde{\boldsymbol{H}}_0$, and $\boldsymbol{k}$ are obtained. Inserting (1.8.7) and (1.8.8) in (1.6.2) and (1.6.3) under the assumption $\tilde{\boldsymbol{J}} = 0$, we find

$$\boldsymbol{k} \times \tilde{\boldsymbol{E}}_0 = \omega\mu\tilde{\boldsymbol{H}}_0, \tag{1.8.9}$$

$$\boldsymbol{k} \times \tilde{\boldsymbol{H}}_0 = -\omega\varepsilon\tilde{\boldsymbol{E}}_0. \tag{1.8.10}$$

Similarly, assuming $\tilde{\rho} = 0$, we get from (1.6.4) and (1.6.5),

$$\boldsymbol{k} \cdot \tilde{\boldsymbol{E}}_0 = 0, \tag{1.8.11}$$

$$\boldsymbol{k} \cdot \tilde{\boldsymbol{H}}_0 = 0. \tag{1.8.12}$$

From (1.8.11) and (1.8.12), we notice that $\tilde{\boldsymbol{E}}_0$ and $\tilde{\boldsymbol{H}}_0$ are perpendicular to $\boldsymbol{k}$. Then, from (1.8.9) and (1.8.10), we find that $\tilde{\boldsymbol{E}}_0$ and $\tilde{\boldsymbol{H}}_0$ are perpendicular to each other, and $\tilde{\boldsymbol{E}}_0$, $\tilde{\boldsymbol{H}}_0$ and $\boldsymbol{k}$ constitute a right-hand system.

We now multiply both sides of (1.8.9) by $\boldsymbol{k}$ vectorially, and use (1.8.10) on the right-hand side of the resultant equation. Then, we get

$$\boldsymbol{k} \times (\boldsymbol{k} \times \tilde{\boldsymbol{E}}_0) = \omega\mu\boldsymbol{k} \times \tilde{\boldsymbol{H}}_0 = -\omega^2\varepsilon\mu\tilde{\boldsymbol{E}}_0. \tag{1.8.13}$$

Developing the left-hand side of (1.8.13) with the aid of the formula for vector triple products, and using the relation (1.8.11), we find

$$(k^2 - \omega^2 \varepsilon\mu)\tilde{\boldsymbol{E}}_0 = 0. \tag{1.8.14}$$

Since $\tilde{\boldsymbol{E}}_0$ does not vanish, we readily get, from (1.8.14), the relation

$$k = \pm\omega\sqrt{\varepsilon\mu} = \pm\frac{\omega}{u}. \tag{1.8.15}$$

The same result as (1.8.15) has already been obtained from (1.8.5) and (1.8.6). From (1.8.9) and (1.8.15), the ratio of the amplitude of electric field to that of magnetic field is found to be

$$\frac{|\tilde{\boldsymbol{E}}_0|}{|\tilde{\boldsymbol{H}}_0|} = \frac{\omega\mu}{|\boldsymbol{k}|} = \sqrt{\frac{\mu}{\varepsilon}}, \tag{1.8.16}$$

which is the intrinsic wave impedance of the medium with permittivity ε and permeability μ.

Finally, let us consider the power flow and the energy stored for the plane wave given by (1.8.3) and (1.8.4). The time-average Poynting vector, or the time-average power flow per unit area for the plane wave is given by

$$\langle \boldsymbol{S} \rangle = \frac{1}{2}\mathrm{Re}[\tilde{\boldsymbol{E}} \times \tilde{\boldsymbol{H}}^*], \tag{1.8.17}$$

which can be calculated with the aid of (1.8.7), (1.8.8), (1.8.9) and (1.8.16) as

$$\langle \boldsymbol{S} \rangle = \frac{1}{2}\boldsymbol{E}_0 \times \boldsymbol{H}_0 = \boldsymbol{i}\frac{1}{2}\sqrt{\frac{\varepsilon}{\mu}}|\boldsymbol{E}_0|^2, \tag{1.8.18}$$

where $\boldsymbol{i}$ denotes a unit vector in the direction of wave propagation.

On the other hand, the time-average electric energy density $\langle w_e \rangle$ and the time-average magnetic energy density $\langle w_m \rangle$ are expressed as

$$\langle w_e \rangle = \frac{1}{4}\varepsilon|\boldsymbol{E}_0|^2, \tag{1.8.19}$$

$$\langle w_m \rangle = \frac{1}{4}\mu|\boldsymbol{H}_0|^2 = \frac{1}{4}\varepsilon|\boldsymbol{E}_0|^2, \tag{1.8.20}$$

from which we find for the total electromagnetic energy density $\langle w \rangle$,

$$\langle w \rangle = \langle w_e \rangle + \langle w_m \rangle = \frac{1}{2}\varepsilon|\boldsymbol{E}_0|^2. \tag{1.8.21}$$

As seen from (1.8.19) and (1.8.20), the plane wave carries an equal amount of electric and magnetic energy. In addition, from (1.8.18) and (1.8.21), we get

$$\frac{\langle \boldsymbol{S} \rangle}{\langle w \rangle} = \boldsymbol{i}\frac{1}{\sqrt{\varepsilon\mu}} = \boldsymbol{i}u\,, \tag{1.8.22}$$

which represents the energy transport velocity in a nondispersive medium.

1.9 Solutions of Inhomogeneous Wave Equations

Let us work out solutions of inhomogeneous wave equations (1.5.10) and (1.5.11) for vector and scalar potentials. For this purpose, we consider an inhomogeneous scalar wave equation of the form,

$$\nabla^2\psi - \frac{1}{u^2}\frac{\partial^2\psi}{\partial t^2} = -f(\boldsymbol{r},t)\,, \tag{1.9.1}$$

where ψ corresponds to a scalar potential ϕ or any component of a vector potential $\boldsymbol{A}$, f a source producing the field ψ, and u the speed of light in an unbounded nondispersive medium with permittivity ε and permeability μ. For simplicity, we seek a solution of the inhomogeneous wave equation (1.9.1) in an unbounded region with the aid of the Fourier transform to the frequency domain. To this end, we assume that $\psi(\boldsymbol{r}, t)$ and $f(\boldsymbol{r}, t)$ have the Fourier integral representations,

$$\psi(\boldsymbol{r},t) = \frac{1}{2\pi}\int_{-\infty}^{\infty}\tilde{\psi}(\boldsymbol{r},\omega)e^{j\omega t}d\omega\,, \tag{1.9.2}$$

$$f(\boldsymbol{r},t) = \frac{1}{2\pi}\int_{-\infty}^{\infty}\tilde{f}(\boldsymbol{r},\omega)e^{j\omega t}d\omega\,, \tag{1.9.3}$$

where,

$$\tilde{\psi}(\boldsymbol{r},\omega) = \int_{-\infty}^{\infty}\psi(\boldsymbol{r},t)e^{-j\omega t}dt\,, \tag{1.9.4}$$

$$\tilde{f}(\boldsymbol{r},\omega) = \int_{-\infty}^{\infty}f(\boldsymbol{r},t)e^{-j\omega t}dt\,. \tag{1.9.5}$$

Inserting the Fourier integral representations (1.9.2) and (1.9.3) into (1.9.1), we find that the Fourier component $\tilde{\psi}(\boldsymbol{r},\omega)$ satisfies the following inhomogeneous Helmholtz equation:

$$(\nabla^2 + k^2)\tilde{\psi}(\boldsymbol{r},\omega) = -\tilde{f}(\boldsymbol{r},\omega)\,, \tag{1.9.6}$$

where $k = \omega/u$ denotes the wave number of the plane wave in an unbounded medium with permittivity ε and permeability μ.

The solution of (1.9.6) can be represented by the weighted superposition of solutions corresponding to a unit point source located at $\boldsymbol{r}'$. The Green function for the inhomogeneous Helmholtz equation (1.9.6), namely, the solution corresponding to a unit point source satisfies the equation

$$\nabla^2 \tilde{G}(\boldsymbol{r},\omega;\boldsymbol{r}') + k^2 \tilde{G}(\boldsymbol{r},\omega;\boldsymbol{r}') = -\delta(\boldsymbol{r}-\boldsymbol{r}'), \tag{1.9.7}$$

with

$$\delta(\boldsymbol{r}-\boldsymbol{r}') = \delta(x-x')\delta(y-y')\delta(z-z'), \tag{1.9.8}$$

where δ is the Dirac delta function. With the aid of the Green function, the Fourier component $\tilde{\psi}(\boldsymbol{r},\omega)$ can be expressed by the weighted superposition

$$\tilde{\psi}(\boldsymbol{r},\omega) = \int_V \tilde{G}(\boldsymbol{r},\omega;\boldsymbol{r}') f(\boldsymbol{r}',\omega) dv, \tag{1.9.9}$$

where the integration is carried out in the volume V in which the source f is distributed.

To find a solution of (1.9.7), we note some general features concerning the solution. First, it is spherically symmetric with the center of symmetry at the source point $\boldsymbol{r}'$, and it behaves like a plane wave for large $\boldsymbol{r}$. In addition, in the limit $\omega \to 0$ or $k \to 0$, it reduces to the solution of Poisson's equation in electrostatics. Hence, if we restrict our discussion only to an outgoing wave, we have for the solution of (1.9.7),

$$\tilde{G}(\boldsymbol{r},\omega;\boldsymbol{r}') = \frac{1}{4\pi|\boldsymbol{r}-\boldsymbol{r}'|} e^{-jk|\boldsymbol{r}-\boldsymbol{r}'|}. \tag{1.9.10}$$

Substituting (1.9.10) into (1.9.9) yields

$$\tilde{\psi}(\boldsymbol{r},\omega) = \frac{1}{4\pi}\int_V \frac{\tilde{f}(\boldsymbol{r}',\omega)}{|\boldsymbol{r}-\boldsymbol{r}'|} e^{-jk|\boldsymbol{r}-\boldsymbol{r}'|} dv. \tag{1.9.11}$$

Then, by inserting (1.9.11) in (1.9.2), the solution of (1.9.1) can be represented as

$$\psi(\boldsymbol{r},t) = \frac{1}{8\pi^2}\iint_V \frac{\tilde{f}(\boldsymbol{r}',\omega)}{|\boldsymbol{r}-\boldsymbol{r}'|} e^{j[\omega t - k|\boldsymbol{r}-\boldsymbol{r}'|]} dv d\omega. \tag{1.9.12}$$

If we use the relation (1.9.3), we finally get

$$\psi(\boldsymbol{r},t) = \frac{1}{4\pi}\int_V \frac{f(\boldsymbol{r}',t')}{|\boldsymbol{r}-\boldsymbol{r}'|} dv, \tag{1.9.13}$$

with

$$t' = t - \frac{|\boldsymbol{r}-\boldsymbol{r}'|}{u}, \tag{1.9.14}$$

where t' denotes the retarded time, which is defined as the time at which the signal arriving at the position $\boldsymbol{r}$ at the time t is emitted at the position $\boldsymbol{r}'$. Note that the signal emitted at the position $\boldsymbol{r}'$ is propagated at the speed u over the distance between the two positions.

From the above discussion, the solutions of the inhomogeneous wave equations (1.5.10) and (1.5.11) for the vector potential $\boldsymbol{A}$ and the scalar potential ϕ can be represented as

$$\boldsymbol{A}(\boldsymbol{r},t) = \frac{\mu}{4\pi}\int_V \frac{\boldsymbol{J}(\boldsymbol{r}',t')}{|\boldsymbol{r}-\boldsymbol{r}'|}dv, \tag{1.9.15}$$

$$\phi(\boldsymbol{r},t) = \frac{1}{4\pi\varepsilon}\int_V \frac{\rho(\boldsymbol{r}',t')}{|\boldsymbol{r}-\boldsymbol{r}'|}dv. \tag{1.9.16}$$

The electric and magnetic fields corresponding to the vector and scalar potentials given above can be calculated from (1.5.1) and (1.5.4).

1.10 Electromagnetic Radiation in Unbounded Space

1.10.1 General Case

In this section, we calculate electromagnetic fields produced by time-varying charge and current distributed in an unbounded medium with permittivity ε and permeability μ with the aid of the Fourier analysis described in the preceding section. The Fourier components or frequency components of the vector and scalar potentials $\boldsymbol{A}(\boldsymbol{r}, t)$ and $\phi(\boldsymbol{r}, t)$ produced by time-varying current and charge densities, $\boldsymbol{J}(\boldsymbol{r}, t)$ and $\rho(\boldsymbol{r}, t)$ are expressed as

$$\tilde{\boldsymbol{A}}(\boldsymbol{r},\omega) = \frac{\mu}{4\pi}\int_V \frac{\tilde{\boldsymbol{J}}(\boldsymbol{r}',\omega)}{|\boldsymbol{r}-\boldsymbol{r}'|}e^{-jk|\boldsymbol{r}-\boldsymbol{r}'|}dv, \tag{1.10.1}$$

$$\tilde{\phi}(\boldsymbol{r},\omega) = \frac{1}{4\pi\varepsilon}\int_V \frac{\tilde{\rho}(\boldsymbol{r}',\omega)}{|\boldsymbol{r}-\boldsymbol{r}'|}e^{-jk|\boldsymbol{r}-\boldsymbol{r}'|}dv, \tag{1.10.2}$$

where $\tilde{\boldsymbol{J}}(\boldsymbol{r}',\omega)$ and $\tilde{\rho}(\boldsymbol{r}',\omega)$ are the Fourier components of $\boldsymbol{J}(\boldsymbol{r}, t)$ and $\rho(\boldsymbol{r}, t)$ defined by

$$\tilde{\boldsymbol{J}}(\boldsymbol{r},\omega) = \int_{-\infty}^{\infty} \boldsymbol{J}(\boldsymbol{r},t)e^{-j\omega t}dt, \tag{1.10.3}$$

$$\tilde{\rho}(\boldsymbol{r},\omega) = \int_{-\infty}^{\infty} \rho(\boldsymbol{r},t)e^{-j\omega t}dt. \tag{1.10.4}$$

The Fourier component of electric field vector, $\tilde{\boldsymbol{E}}(\boldsymbol{r},\omega)$ can be evaluated from the relation

$$\tilde{\boldsymbol{E}}(\boldsymbol{r},\omega) = -\nabla\tilde{\phi}(\boldsymbol{r},\omega) - j\omega\tilde{\boldsymbol{A}}(\boldsymbol{r},\omega). \tag{1.10.5}$$

Substituting (1.10.1) and (1.10.2) into (1.10.5), we get

$$\tilde{\boldsymbol{E}}(\boldsymbol{r},\omega) = -\frac{1}{4\pi\varepsilon}\nabla\left(\int_V \frac{\tilde{\rho}(\boldsymbol{r}',\omega)}{|\boldsymbol{r}-\boldsymbol{r}'|}e^{-jk|\boldsymbol{r}-\boldsymbol{r}'|}dv\right)$$

$$-\frac{j\omega\mu}{4\pi}\int_V \frac{\tilde{\boldsymbol{J}}(\boldsymbol{r}',\omega)}{|\boldsymbol{r}-\boldsymbol{r}'|}e^{-jk|\boldsymbol{r}-\boldsymbol{r}'|}dv. \tag{1.10.6}$$

To cast the right-hand side of (1.10.6) in a more convenient form, some manipulations are required [1.2]. Interchanging the order of differentiation and integration in the first term on the right-hand side of (1.10.6), we find

$$\tilde{\boldsymbol{E}} = \frac{1}{4\pi\varepsilon}\int_V \boldsymbol{i}_R\tilde{\rho}\frac{e^{-jk|\boldsymbol{r}-\boldsymbol{r}'|}}{|\boldsymbol{r}-\boldsymbol{r}'|^2}dv$$

$$+\frac{jk}{4\pi\varepsilon}\int_V\left(\boldsymbol{i}_R\tilde{\rho}-\frac{\tilde{\boldsymbol{J}}}{u}\right)\frac{e^{-jk|\boldsymbol{r}-\boldsymbol{r}'|}}{|\boldsymbol{r}-\boldsymbol{r}'|}dv, \tag{1.10.7}$$

where we have omitted arguments attached to the Fourier components, and $\boldsymbol{i}_R$ denotes a unit vector directed from the source point $\boldsymbol{r}'$ to the field point $\boldsymbol{r}$,

$$\boldsymbol{i}_R = \frac{\boldsymbol{r}-\boldsymbol{r}'}{|\boldsymbol{r}-\boldsymbol{r}'|}. \tag{1.10.8}$$

Next, let us rewrite the second term on the right-hand side of (1.10.7). To deform this integral, which will be denoted by $\tilde{\boldsymbol{E}}_2$, we use the law of charge conservation, (1.6.6). Then, we have

$$\tilde{\boldsymbol{E}}_2 = -\frac{1}{4\pi\varepsilon u}\int_V\left(\boldsymbol{i}_R(\nabla'\cdot\tilde{\boldsymbol{J}}) + jk\tilde{\boldsymbol{J}}\right)\frac{e^{-jk|\boldsymbol{r}-\boldsymbol{r}'|}}{|\boldsymbol{r}-\boldsymbol{r}'|}dv, \tag{1.10.9}$$

where ∇' represents the differential operator with respect to the source point. We can evaluate (1.10.9) by decomposing it into rectangular components and applying partial integration. After some manipulation, we get

$$\tilde{\boldsymbol{E}}_2 = \frac{1}{4\pi\varepsilon u}\int_V\left[2(\tilde{\boldsymbol{J}}\cdot\boldsymbol{i}_R)\boldsymbol{i}_R - \tilde{\boldsymbol{J}}\right]\frac{e^{-jk|\boldsymbol{r}-\boldsymbol{r}'|}}{|\boldsymbol{r}-\boldsymbol{r}'|^2}dv$$

$$+\frac{jk}{4\pi\varepsilon u}\int_V\left[(\tilde{\boldsymbol{J}}\times\boldsymbol{i}_R)\times\boldsymbol{i}_R\right]\frac{e^{-jk|\boldsymbol{r}-\boldsymbol{r}'|}}{|\boldsymbol{r}-\boldsymbol{r}'|}dv. \tag{1.10.10}$$

From (1.10.7) and (1.10.10), we obtain a final form for the Fourier component of electric field,

$$\tilde{\boldsymbol{E}} = \frac{1}{4\pi\varepsilon}\int_V \boldsymbol{i}_R\tilde{\rho}\frac{e^{-jk|\boldsymbol{r}-\boldsymbol{r}'|}}{|\boldsymbol{r}-\boldsymbol{r}'|^2}dv$$

$$+\frac{1}{4\pi\varepsilon u}\int_V\left[2(\tilde{\boldsymbol{J}}\cdot\boldsymbol{i}_R)\boldsymbol{i}_R-\tilde{\boldsymbol{J}}\right]\frac{e^{-jk|\boldsymbol{r}-\boldsymbol{r}'|}}{|\boldsymbol{r}-\boldsymbol{r}'|^2}dv$$

$$+\frac{jk}{4\pi\varepsilon u}\int_V\left[(\tilde{\boldsymbol{J}}\times\boldsymbol{i}_R)\times\boldsymbol{i}_R\right]\frac{e^{-jk|\boldsymbol{r}-\boldsymbol{r}'|}}{|\boldsymbol{r}-\boldsymbol{r}'|}dv. \qquad (1.10.11)$$

The electric field vector in time domain, $\boldsymbol{E}(\boldsymbol{r}, t)$ corresponding to the counterpart in frequency domain, $\tilde{\boldsymbol{E}}(\boldsymbol{r},\omega)$ can be obtained from the inverse Fourier transform,

$$\boldsymbol{E}(\boldsymbol{r},t) = \frac{1}{2\pi}\int_{-\infty}^{\infty}\tilde{\boldsymbol{E}}(\boldsymbol{r},\omega)e^{j\omega t}d\omega. \qquad (1.10.12)$$

Applying (1.10.11) to (1.10.12), we get for the electric field vector in time domain,

$$\boldsymbol{E}(\boldsymbol{r},t) = \frac{1}{4\pi\varepsilon}\int_V \boldsymbol{i}_R\frac{\rho(r',t')}{|\boldsymbol{r}-\boldsymbol{r}'|^2}dv$$

$$+\frac{1}{4\pi\varepsilon u}\int_V\frac{2\left[\boldsymbol{J}(\boldsymbol{r}',t')\cdot\boldsymbol{i}_R\right]\boldsymbol{i}_R-\boldsymbol{J}(\boldsymbol{r}',t')}{|\boldsymbol{r}-\boldsymbol{r}'|^2}dv$$

$$+\frac{\mu}{4\pi}\int_V\frac{\left[\dot{\boldsymbol{J}}(\boldsymbol{r}',t')\times\boldsymbol{i}_R\right]\times\boldsymbol{i}_R}{|\boldsymbol{r}-\boldsymbol{r}'|}dv, \qquad (1.10.13)$$

with

$$\dot{\boldsymbol{J}}(\boldsymbol{r}',t') = \frac{\partial}{\partial t'}\boldsymbol{J}(\boldsymbol{r}',t') = \frac{1}{2\pi}\int_{-\infty}^{\infty}j\omega\tilde{\boldsymbol{J}}(\boldsymbol{r}',\omega)e^{j\omega t'}d\omega, \qquad (1.10.14)$$

where it should be noted that the charge and current densities and the time derivative of the current density are evaluated at the retarded time t'.

The Fourier component of magnetic field vector, $\tilde{\boldsymbol{H}}(\boldsymbol{r},\omega)$ can be calculated from

$$\tilde{\boldsymbol{H}}(\boldsymbol{r},\omega) = \frac{1}{\mu}\tilde{\boldsymbol{B}}(\boldsymbol{r},\omega) = \frac{1}{\mu}\nabla\times\tilde{\boldsymbol{A}}(\boldsymbol{r},\omega). \qquad (1.10.15)$$

Inserting (1.10.1) in (1.10.15), we find

$$\tilde{\boldsymbol{H}} = \frac{1}{4\pi}\nabla\times\int_V\frac{\tilde{\boldsymbol{J}}(\boldsymbol{r}',\omega)}{|\boldsymbol{r}-\boldsymbol{r}'|}e^{-jk|\boldsymbol{r}-\boldsymbol{r}'|}dv$$

$$= -\frac{1}{4\pi}\int_V \tilde{\boldsymbol{J}}(r',\omega)\times\nabla\left(\frac{e^{-jk|\boldsymbol{r}-\boldsymbol{r}'|}}{|\boldsymbol{r}-\boldsymbol{r}'|}\right)dv$$

$$= \frac{1}{4\pi}\left[\int_V \frac{\tilde{\boldsymbol{J}}(\boldsymbol{r}',\omega)\times\boldsymbol{i}_R}{|\boldsymbol{r}-\boldsymbol{r}'|^2}e^{-jk|\boldsymbol{r}-\boldsymbol{r}'|}dv + jk\int_V \frac{\tilde{\boldsymbol{J}}(\boldsymbol{r}',\omega)\times\boldsymbol{i}_R}{|\boldsymbol{r}-\boldsymbol{r}'|}e^{-jk|\boldsymbol{r}-\boldsymbol{r}'|}dv\right].$$

(1.10.16)

Applying the inverse Fourier transform to (1.10.16), we get for the magnetic field vector in time domain,

$$\boldsymbol{H}(\boldsymbol{r},t) = \frac{1}{2\pi}\int_{-\infty}^{\infty}\tilde{\boldsymbol{H}}(\boldsymbol{r},\omega)e^{j\omega t}d\omega$$

$$= \frac{1}{4\pi}\int_V\left(\frac{\boldsymbol{J}(\boldsymbol{r}',t')\times\boldsymbol{i}_R}{|\boldsymbol{r}-\boldsymbol{r}'|^2} + \frac{1}{u}\frac{\dot{\boldsymbol{J}}(\boldsymbol{r}',t')\times\boldsymbol{i}_R}{|\boldsymbol{r}-\boldsymbol{r}'|}\right)dv. \quad (1.10.17)$$

Extracting the radiation fields alone from (1.10.13) and (1.10.17), we have

$$\boldsymbol{E}_{rad} = \frac{\mu}{4\pi}\int_V \frac{\left[\dot{\boldsymbol{J}}(\boldsymbol{r}',t')\times\boldsymbol{i}_R\right]\times\boldsymbol{i}_R}{|\boldsymbol{r}-\boldsymbol{r}'|}dv, \quad (1.10.18)$$

$$\boldsymbol{H}_{rad} = \frac{1}{4\pi u}\int_V \frac{\dot{\boldsymbol{J}}(\boldsymbol{r}',t')\times\boldsymbol{i}_R}{|\boldsymbol{r}-\boldsymbol{r}'|}dv. \quad (1.10.19)$$

At large distances from the charge and current distributions, the radiation fields predominate in the total fields given by (1.10.13) and (1.10.17). For their main features, they decay in inverse proportion to the distance from the origin around which the charge and current are distributed, thus contributing to radiation of electromagnetic energy. In addition, they are transverse to the direction of propagation, and behave like plane waves.

1.10.2 Electric Dipole Radiation

For a typical example of electromagnetic radiation, let us consider the radiation from an oscillating electric dipole. For this purpose, let a sinusoidally time-varying current be uniformly flowing along a thin short wire of length h placed on the z-axis with its center coinciding with the origin. Specifically, the current density for the current specified above is represented mathematically in frequency domain as

$$\tilde{\boldsymbol{J}}(r',\omega) = \boldsymbol{i}_z I_0\delta(x')\delta(y'),\ \left(-\frac{h}{2} < z' < \frac{h}{2}\right), \quad (1.10.20)$$

where I_0 denotes the magnitude of a sinusoidally time-varying current. According to the law of charge conservation, time-varying charges appear at both ends of the wire. The charge density corresponding to these charges in frequency domain is expressed as

$$\tilde{\rho}(\boldsymbol{r}',\omega) = \frac{I_0}{j\omega}\delta(x')\delta(y')\left[\delta(z' - \frac{h}{2}) - \delta(z' + \frac{h}{2})\right]. \tag{1.10.21}$$

For the current and charge densities given by (1.10.20) and (1.10.21) for a thin short wire, the electric and magnetic fields in frequency domain given by (1.10.11) and (1.10.16) are simplified. First, to rewrite the first term on the right-hand side of (1.10.11), we note some approximating relations. At both ends of the wire, we have an approximation for the phase term

$$e^{-jk|\boldsymbol{r}-\boldsymbol{r}'|} = e^{-jkr}\left[1 \pm jk\frac{h}{2}\cos\theta\right]. \tag{1.10.22}$$

In addition, for the unit vector $\boldsymbol{i}_R$ directed from the source point to the field point, we note the following relation:

$$\boldsymbol{i}_R = \boldsymbol{i}_r \pm \boldsymbol{i}_\theta \frac{h}{2r}\sin\theta, \tag{1.10.23}$$

where $\boldsymbol{i}_r$ and $\boldsymbol{i}_\theta$ are unit vectors in the r and θ directions, respectively. In (1.10.22) and (1.10.23), the upper sign corresponds to $z' = h/2$ and the lower sign to $z' = -h/2$.

Taking the relations (1.10.22) and (1.10.23) into account, we can rewrite the first term on the right-hand side of (1.10.11) as

$$\frac{1}{4\pi\varepsilon}\int_V \boldsymbol{i}_R\tilde{\rho}\frac{e^{-jk|\boldsymbol{r}-\boldsymbol{r}'|}}{|\boldsymbol{r}-\boldsymbol{r}'|^2}dv = \frac{I_0 h}{4\pi}e^{-jkr}\left(\frac{1}{j\omega\varepsilon r^3}\right)\left(\boldsymbol{i}_r 2\cos\theta + \boldsymbol{i}_\theta\sin\theta\right)$$

$$+\boldsymbol{i}_r\frac{I_0 h}{4\pi}e^{-jkr}\sqrt{\frac{\mu}{\varepsilon}}\frac{1}{r^2}\cos\theta. \tag{1.10.24}$$

Similarly, we can reduce the other terms of the electric and magnetic fields in frequency domain, (1.10.11) and (1.10.16).

The final results for the electric and magnetic field components in spherical coordinates are obtained as follows:

$$\tilde{E}_r = \frac{I_0 hk^3}{2\pi\omega\varepsilon}e^{-jkr}\left[\frac{1}{(kr)^2} - \frac{j}{(kr)^3}\right]\cos\theta, \tag{1.10.25}$$

$$\tilde{E}_\theta = \frac{I_0 hk^3}{4\pi\omega\varepsilon}e^{-jkr}\left[\frac{j}{kr} + \frac{1}{(kr)^2} - \frac{j}{(kr)^3}\right]\sin\theta, \tag{1.10.26}$$

$$\tilde{H}_\varphi = \frac{I_0 hk^2}{4\pi}e^{-jkr}\left[\frac{j}{kr} + \frac{1}{(kr)^2}\right]\sin\theta, \tag{1.10.27}$$

$$\tilde{E}_\varphi = \tilde{H}_r = \tilde{H}_\theta = 0. \tag{1.10.28}$$

As stated in the preceding subsection, the radiation fields, which decay in inverse proportion to the distance from the origin, predominate at large distances ($kr >>1$) from the origin where an electric dipole is located. In the vicinity of an electric dipole ($kr <<1$), on the other hand, the electric and magnetic fields behave in a different manner. Specifically, the electric field varies in inverse proportion to the cube of the distance from the origin, while the magnetic field varies in inverse proportion to the square of the distance from the origin. In the vicinity of the origin, the electric field pattern is the same as produced by an electrostatic dipole, whereas the magnetic field pattern is the same as predicted by Ampere's law.

1.10.3 Radiated Energy and Power

Let us first consider the radiation from a pulse source of finite duration. At sufficiently large distances from the source, the radiation fields in time domain given by (1.10.18) and (1.10.19) are simplified as

$$\boldsymbol{E}_{rad} = \sqrt{\frac{\mu}{\varepsilon}}\boldsymbol{H}_{rad} \times \boldsymbol{i}_r, \tag{1.10.29}$$

$$\boldsymbol{H}_{rad} = \frac{1}{4\pi u r}\int_V \left(\dot{\boldsymbol{J}} \times \boldsymbol{i}_r\right) dv. \tag{1.10.30}$$

The total power P radiated from a pulse source at an arbitrary time is found by integrating the Poynting vector associated with the radiation fields over a sufficiently large sphere of radius r.

$$\begin{aligned} P &= \oint_S \left(\boldsymbol{E}_{rad} \times \boldsymbol{H}_{rad}\right)\cdot \boldsymbol{i}_r ds \\ &= \frac{1}{16\pi^2 u^2}\sqrt{\frac{\mu}{\varepsilon}}\oint_\Omega \left|\int_V \left(\dot{\boldsymbol{J}} \times \boldsymbol{i}_r\right) dv\right|^2 d\Omega, \end{aligned} \tag{1.10.31}$$

where $d\Omega$ denotes the solid angle subtended by ds at the origin.

The total energy U radiated by the pulse source is obtained by integrating (1.10.31) over the pulse duration,

$$U = \int_{-\tau}^{\tau} P dt = \int_{-\infty}^{\infty} P dt, \tag{1.10.32}$$

where 2τ is the time of pulse duration. The total energy radiated by a pulse source can also be evaluated in frequency domain with the aid of the Fourier transform. For this purpose, we note Parseval's theorem in the Fourier transform. If we are given a temporal function $f(t)$ for which there exists the Fourier transform $\tilde{f}(\omega)$,

Parseval's theorem reads as follows:

$$\int_{-\infty}^{\infty} |f(t)|^2 dt = \frac{1}{\pi}\int_0^{\infty} \left|\tilde{f}(\omega)\right|^2 d\omega, \tag{1.10.33}$$

where the factor $1/\pi$ depends on the factor appearing in the definition of the Fourier transform. Note that we defined the Fourier transform as in (1.9.2) and (1.9.4).

To calculate the total energy radiated from a pulse source in frequency domain, we simplify the Fourier components for the radiation fields given by (1.10.11) and (1.10.16). In the integrands of these equations, we can apply the same approximations for amplitude factors as used for obtaining (1.10.29) and (1.10.30). On the other hand, for the phase terms in the integrands of (1.10.11) and (1.10.16), we must use the approximation

$$k|\boldsymbol{r} - \boldsymbol{r}'| = kr - \boldsymbol{k}\cdot\boldsymbol{r}', \tag{1.10.34}$$

with

$$\boldsymbol{k} = \boldsymbol{i}_r k. \tag{1.10.35}$$

Then we get for the approximate expressions for the radiation fields in frequency domain,

$$\tilde{\boldsymbol{E}}_{rad} = \frac{1}{\omega\varepsilon}\tilde{\boldsymbol{H}}_{rad} \times \boldsymbol{k}, \tag{1.10.36}$$

$$\tilde{\boldsymbol{H}}_{rad} = \frac{j}{4\pi r} e^{-jkr}\int_V \left(\tilde{\boldsymbol{J}} \times \boldsymbol{k}\right) e^{j\boldsymbol{k}\cdot\boldsymbol{r}'} dv. \tag{1.10.37}$$

With the aid of (1.10.33), together with (1.10.36) and (1.10.37), we obtain the total energy radiated by a pulse source

$$U = \frac{1}{16\pi^3}\sqrt{\frac{\mu}{\varepsilon}}\oint_{\Omega}\int_0^{\infty}\left|\int_V \left(\tilde{\boldsymbol{J}} \times \boldsymbol{k}\right) e^{j\boldsymbol{k}\cdot\boldsymbol{r}'} dv\right|^2 d\omega d\Omega, \tag{1.10.38}$$

from which we find for the energy radiated per unit solid angle,

$$\frac{dU}{d\Omega} = \frac{1}{16\pi^3}\sqrt{\frac{\mu}{\varepsilon}}\int_0^{\infty}\left|\int_V \left(\tilde{\boldsymbol{J}} \times \boldsymbol{k}\right) e^{j\boldsymbol{k}\cdot\boldsymbol{r}'} dv\right|^2 d\omega. \tag{1.10.39}$$

For a monochromatic source oscillating with a single frequency ω, the total energy radiated becomes infinite. Thus for this case, it is appropriate to consider the time-average energy radiated per second or the time-average radiated power. Note that the time-average Poynting vector for a monochromatic source is defined as

$$\langle \boldsymbol{S} \rangle = \frac{1}{2}\mathrm{Re}\left[\tilde{\boldsymbol{E}}_{rad} \times \tilde{\boldsymbol{H}}^*_{rad}\right]. \tag{1.10.40}$$

Then we have for the time-average total power radiated from a monochromatic source,

$$\langle P \rangle = \oint_S \langle \boldsymbol{S} \rangle \cdot \boldsymbol{i}_r ds$$
$$= \frac{1}{32\pi^2} \sqrt{\frac{\mu}{\varepsilon}} \oint_\Omega \left| \int_V \left(\tilde{\boldsymbol{J}} \times \boldsymbol{k} \right) e^{j\boldsymbol{k}\cdot\boldsymbol{r}'} dv \right|^2 d\Omega. \tag{1.10.41}$$

From (1.10.41), the time-average power radiated per unit solid angle is given by

$$\frac{d\langle P \rangle}{d\Omega} = \frac{1}{32\pi^2} \sqrt{\frac{\mu}{\varepsilon}} \left| \int_V \left(\tilde{\boldsymbol{J}} \times \boldsymbol{k} \right) e^{j\boldsymbol{k}\cdot\boldsymbol{r}'} dv \right|^2. \tag{1.10.42}$$

For a specific example, let us consider the radiation from an electric dipole oscillating with single frequency ω. The total average power radiated from an oscillating electric dipole is found from (1.10.26) and (1.10.27) as

$$\langle P \rangle = \frac{(I_0 hk)^2}{12\pi} \sqrt{\frac{\mu}{\varepsilon}}, \tag{1.10.43}$$

while the time-average power radiated per unit solid angle is given by

$$\frac{d\langle P \rangle}{d\Omega} = \frac{(I_0 hk)^2}{32\pi^2} \sqrt{\frac{\mu}{\varepsilon}} \sin^2\theta. \tag{1.10.44}$$

2. Foundations of Relativistic Electrodynamics

2.1 Special Theory of Relativity

In this chapter, we describe the foundations of electromagnetic theory as viewed from the standpoint of the special theory of relativity. The special theory of relativity gives the most basic foundation that underlies all physical laws in inertial systems of reference. Inertial systems of reference are those systems of reference for which the law of inertia holds. According to the law of inertia or Newton's first law, a material particle in an inertial system of reference, which is free from external influences, will stay at rest or continue to move in a straight line with constant velocity. Before we start to discuss the essential parts of relativistic electromagnetic theory, we briefly summarize the historical background from which the special theory of relativity emerged.

The second law of the Newtonian mechanics, which had occupied the absolute position in physics before the latter half of the 19th century, is a basic law of motion which holds in an inertial system of reference. However, any system of reference moving with uniform velocity relative to an inertial system of reference is also another inertial system of reference, because the law of inertia holds in the latter system. Thus it follows that there are infinite sets of inertial systems of reference and Newton's second law holds in all these inertial systems. This principle is called the Galilean principle of relativity. In other words, the form of Newton's second law is kept invariant under the coordinate transformation referred to as the Galilean transformation. For the later discussion, it should be noted that the usual addition theorem of velocities holds for the Galilean transformation.

On the other hand, the basic law to describe electromagnetic phenomena is the Maxwell equations established by Maxwell in 1864. With the aid of his equations, he predicted the existence of electromagnetic waves, showing that the light is a kind of electromagnetic wave. According to the wave theory of light at that time, the light was considered to propagate in a kind of medium called the ether. Hence, when viewed from an inertial system of reference that is at rest relative to the ether, the light should have propagated uniformly in every direction with the velocity of light c. On the other hand, when viewed from an inertial system moving uniformly relative to the ether, the light should have traveled with different velocities in different directions. On the basis of this deduction, Michelson, Morley, and other experimental physicists carried out various optical experiments to detect the translational motion of the earth relative to the ether, but all these experiments gave negative results. Thus, according to the experimental results, it followed that the light travels uniformly with constant velocity c in every direction on the earth as well, which was assumed to be moving relative to the ether. This result shows that the usual addition theorem of velocities for the Galilean transformation does not apparently hold for the electromagnetic phenomenon of light propagation. At the same time, it suggests that the concept of the ether must be discarded.

The difficulties in physics concerning the ether and inertial systems of reference, which were revealed by the experiments of Michelson-Morley and others, were resolved with the advent of the special theory of relativity established by Einstein in 1905. The special theory of relativity is built on two postulates: one is *the special principle of relativity*, and the other is *the constancy of the light velocity in inertial systems of reference*. According to the special principle of relativity, all physical laws should be formulated in the same form in all inertial systems of reference. In other words, one cannot, in principle, select any one special system or the absolute system to describe physical phenomena in inertial systems of reference. Instead, one can regard all inertial systems as equivalent for describing physical phenomena. On the other hand, the latter postulate of the constancy of the light velocity in inertial systems means that the light travels with constant velocity c in every direction in all inertial systems set in vacuum, independently of the state of motion of the light source. According to the latter postulate, one can judge if a particular system of reference is an inertial system, by measuring the velocity of light in that system of reference. On the basis of the aforementioned two postulates, Einstein derived the Lorentz transformation for a new coordinate transformation connecting two inertial systems of reference, which replaced the Galilean transformation.

2.2 Lorentz Transformations

Let us derive a new coordinate transformation from one inertial system to another, which replaces the Galilean transformation, on the basis of Einstein's two postulates for the special theory of relativity. For this purpose, we consider two inertial systems $I(x, y, z, t)$ and $I'(x', y', z', t')$. The inertial system I' is assumed to be uniformly moving in the z direction with constant velocity v relative to the inertial system I, keeping each coordinate axis parallel to the corresponding axis of the latter, as shown in Fig. 2.1. Now, let an event occur at the position (x, y, z) at the time t in the inertial system I, and let the same event occur at the corresponding position (x', y', z') and at the corresponding time t' in the inertial system I'. Then, we try to find how the sets of space-time coordinates, (x, y, z, t) and (x', y', z', t') are transformed to each other under the two postulates for the special theory of relativity. From the law of inertia, it is apparent that a uniform rectilinear motion in one inertial system corresponds to another uniform rectilinear motion in the other

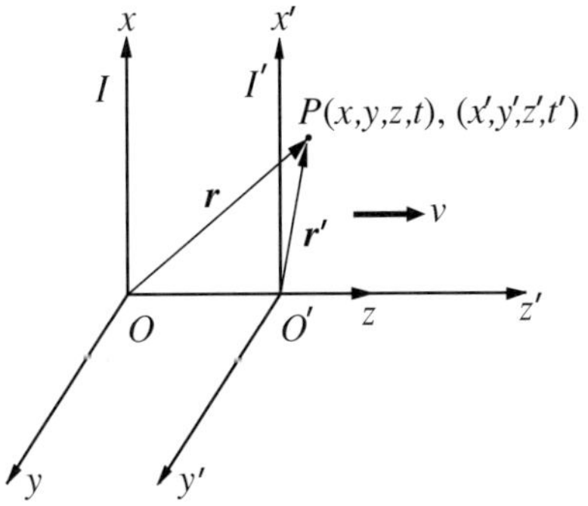

Fig. 2.1. Inertial systems I and I'

inertial system. Hence, first of all, the transformation between two inertial systems I and I' must be given by linear equations in terms of space-time coordinates.

Let the origins of the inertial systems I and I' coincide with each other at the time $t = t' = 0$. Then, the origin of I', namely, $x' = 0$, $y' = 0$, and $z' = 0$ corresponds to $x = 0$, $y = 0$, and $z = vt$. Thus, between (x', y', z') and (x, y, z), we must have the relation

$$x' = \eta(v)x, \quad y' = \eta(v)y, \quad z' = \gamma(v)(z - vt), \tag{2.2.1}$$

where we use the same coefficient $\eta(v)$ before x and y because of the spatial symmetry. Viewed in turn from the inertial system I', the inertial system I is moving uniformly in the z' direction with constant velocity $-v$ relative to the former. Thus, in place of (2.2.1), we have

$$x = \eta(-v)x', \quad y = \eta(-v)y', \quad z = \gamma(-v)(z' + vt'). \tag{2.2.2}$$

We now examine the physical meaning of the coefficients, $\gamma(v)$, $\gamma(-v)$, $\eta(v)$, and $\eta(-v)$ appearing in (2.2.1) and (2.2.2). For this purpose, let a measuring rod with length l_0 be lying on the z'-axis at rest in the inertial system I'. Viewing this measuring rod from the inertial system I, we find the corresponding length of the rod, l from the third equation of (2.2.1),

$$l = \frac{l_0}{\gamma(v)}. \tag{2.2.3}$$

When applying the third equation of (2.2.1) to get the relation (2.2.3), note that we must specify the z coordinates of both ends of the rod at the same instant of time. Equation (2.2.3) shows that the length of the rod viewed from I is reduced to $1/\gamma(v)$ of the proper length l_0. On the other hand, viewing a rod with length l_0 lying along the z-axis, which is at rest in the inertial system I, from the inertial system I', we get for the length of the rod in the system I' from the third equation of (2.2.2),

$$l' = \frac{l_0}{\gamma(-v)}. \tag{2.2.4}$$

Since the situations in (2.2.3) and (2.2.4) are symmetrical, we find the relation $l = l'$ or $\gamma(v) = \gamma(-v)$ from the special principle of relativity. Namely, the coefficient $\gamma(v)$ proves to be an even function of v. From a similar argument, the coefficient $\eta(v)$ is also found to be an even function of v. Thus we set $\gamma(v) = \gamma(-v) = \gamma$, and $\eta(v) = \eta(-v) = \eta$.

First, multiplying both sides of the first equations of (2.2.1) and (2.2.2), we get $\eta^2 = 1$, from which $\eta = 1$ is obtained because two inertial systems coincide with each other when $v = 0$.

Next, with the aid of the constancy of the light velocity in inertial systems, let us try to find the value of γ from the third equations of (2.2.1) and (2.2.2). To this end, let a light pulse be emitted from the origins of the systems I and I' at the time $t' = t = 0$, and let it reach the point $(0,0,z')$ at the time t' in the system I', and the corresponding point $(0,0,z)$ at the corresponding time t in the system I. Then, from the constancy of the light velocity in inertial systems, we have the relations $z' = ct'$ and $z = ct$, where c denotes the velocity of light in vacuum. Inserting these relations into the third equations of (2.2.1) and (2.2.2), we obtain for the value of γ,

$$\gamma = \frac{1}{\sqrt{1-\beta^2}}, \quad \beta = \frac{v}{c}. \tag{2.2.5}$$

Using the values of η and γ obtained above in the third equations of (2.2.1) and (2.2.2), we get the relations between the sets (z',t') and (z,t),

$$z' = \gamma(z - vt), \quad z = \gamma(z' + vt'). \tag{2.2.6}$$

Solving (2.2.6) for t' and t, we find

$$t' = \gamma(t - \frac{v}{c^2} z), \quad t = \gamma(t' + \frac{v}{c^2} z'). \tag{2.2.7}$$

From the above derivation, we finally have for a new space-time transformation between the inertial systems I' and I,

$$x' = x, \quad y' = y, \quad z' = \gamma(z - vt), \quad t' = \gamma(t - \frac{v}{c^2} z), \tag{2.2.8}$$

and

$$x = x', \quad y = y', \quad z = \gamma(z' + vt'), \quad t = \gamma(t' + \frac{v}{c^2} z'). \tag{2.2.9}$$

It should be noted that the transformation (2.2.9) can readily be obtained by replacing the primed space-time coordinates by the unprimed ones, and v by $-v$ in (2.2.8). The space-time transformations (2.2.8) and (2.2.9) are referred to as the

Lorentz transformations. The above space-time transformations were named the Lorentz transformations after Lorentz, who first showed before the advent of the special theory of relativity that the form of the Maxwell equations is kept invariant under the same transformations as given by (2.2.8) and (2.2.9). However, he had to presume the concepts of the Lorentz contraction and the fictitious local time whose physical meanings were not clear at all. Thus he could not reach the special theory of relativity. If we have c tend to infinity in (2.2.8) and (2.2.9), we get the Galilean transformations for which the absolute time is retained.

In the above discussion, we derived the Lorentz transformations for a special case where two inertial systems are moving uniformly relative to each other along the z and z' axes, keeping each coordinate axis in one inertial system parallel to each corresponding axis in the other inertial system. For a general case where the inertial system I' is moving in an arbitrary direction denoted by vector $\boldsymbol{v}$ relative to the inertial system I, the Lorentz transformation from the inertial system I to the inertial system I' is given by the following expressions:

$$\boldsymbol{r}'_{\perp} = \boldsymbol{r}_{\perp}, \quad \boldsymbol{r}'_{//} = \gamma(\boldsymbol{r}_{//} - \boldsymbol{v}t), \quad t' = \gamma\left(t - \frac{1}{c^2}\boldsymbol{v}\cdot\boldsymbol{r}\right), \tag{2.2.10}$$

where $\boldsymbol{r}$ denotes a position vector, and the subscripts $\perp$ and $//$ mean the components of a position vector normal and parallel to velocity vector $\boldsymbol{v}$. The corresponding transformation from the system I' to the system I can be found by replacing the primed quantities by the unprimed quantities, and $\boldsymbol{v}$ by $-\boldsymbol{v}$ in (2.2.10). As is evident from (2.2.8) ~ (2.2.10), the space-time coordinates are combined with each other in relativity.

Finally, we note the physical meaning of the velocity of light c appearing in the Lorentz transformation. The light velocity c in the Lorentz transformation refers to the maximum velocity with which electromagnetic disturbances or interactions spread out. Thus, in material media as well where phase velocities and group velocities are different from the velocity of light in vacuum, c, we can use (2.2.8) ~ (2.2.10) for the Lorentz transformations. This is because in these media as well the maximum velocity for electromagnetic disturbances or interactions to propagate becomes equal to the velocity of light in vacuum. For example, it is well known that the precursor of an electromagnetic pulse is propagated with the velocity of light in vacuum, c in these media as well.

2.3 Transformation of Electromagnetic Quantities

According to the special theory of relativity, the Maxwell equations hold in all inertial systems. In other words, when we move from one inertial system to another by the Lorentz transformation, the form of the Maxwell equations is kept invariant. If a physical law is form-invariant under a transformation as in the above example, the physical law is referred to as covariant under the transformation. Hence the Maxwell equations are covariant under the Lorentz transformation. With the aid of the covariance of the Maxwell equations under the Lorentz transformation, let us try to find the transformation formulas for field vectors, and those for charge and current densities under the Lorentz transformation.

Let an inertial system I' be moving uniformly in the z direction with velocity v relative to another inertial system I. The Maxwell equations in the inertial system I are expressed, from (1.2.8) ~ (1.2.11), as

$$\nabla \times \boldsymbol{E} = -\frac{\partial \boldsymbol{B}}{\partial t}, \quad \nabla \cdot \boldsymbol{B} = 0, \tag{2.3.1}$$

$$\nabla \times \boldsymbol{H} = \frac{\partial \boldsymbol{D}}{\partial t} + \boldsymbol{J}, \quad \nabla \cdot \boldsymbol{D} = \rho. \tag{2.3.2}$$

Equations (2.3.1) and (2.3.2) are arranged differently from (1.2.8) ~ (1.2.11). This is because each of the first pair of equations, (2.3.1) and the second pair of equations, (2.3.2) constitutes an independent law in relativistic electromagnetic theory.

On the other hand, according to the special principle of relativity, the equations of the same form as (2.3.1) and (2.3.2) hold in the inertial system I' as well. Namely, we have in the system I',

$$\nabla' \times \boldsymbol{E}' = -\frac{\partial \boldsymbol{B}'}{\partial t'}, \quad \nabla' \cdot \boldsymbol{B}' = 0, \tag{2.3.3}$$

$$\nabla' \times \boldsymbol{H}' = \frac{\partial \boldsymbol{D}'}{\partial t'} + \boldsymbol{J}', \quad \nabla' \cdot \boldsymbol{D}' = \rho', \tag{2.3.4}$$

where the primes denote the electromagnetic quantities and the operations in the system I'. The transformation formulas for field vectors and those for charge and current densities are obtained from the requirement that we should have the Maxwell equations in the system I', (2.3.3) and (2.3.4) when we transform those in the system I, (2.3.1) and (2.3.2) to the system I' by the Lorentz transformation.

In order to transform Eqs. (2.3.1) and (2.3.2) to the system I', let us first try to find the transformation formulas for the spatial and temporal derivatives under the Lorentz transformation (2.2.8). For this purpose, let $f(x, y, z, t)$ be an arbitrary scalar function of variables x, y, z, t. Then, the function $f(x, y, z, t)$ can be regarded as a function of variables x', y', z', t' as well, since variables x, y, z, t are related to variables x', y', z', t' through (2.2.8). Hence, for example, the derivative of $f(x, y, z, t)$ with respect to z can be represented, with the aid of the differentiation formula and (2.2.8), as

$$\begin{aligned}\frac{\partial f}{\partial z} &= \frac{\partial f}{\partial z'}\frac{\partial z'}{\partial z} + \frac{\partial f}{\partial t'}\frac{\partial t'}{\partial z} \\ &= \gamma \frac{\partial f}{\partial z'} - \gamma \frac{v}{c^2}\frac{\partial f}{\partial t'}\end{aligned}$$

$$= \gamma\left(\frac{\partial}{\partial z'} - \frac{v}{c^2}\frac{\partial}{\partial t'}\right)f. \tag{2.3.5}$$

In a similar fashion, we can get for the derivatives of $f(x, y, z, t)$ with respect to x, y, and t,

$$\frac{\partial f}{\partial x} = \frac{\partial f}{\partial x'}, \quad \frac{\partial f}{\partial y} = \frac{\partial f}{\partial y'}, \quad \frac{\partial f}{\partial t} = \gamma\left(\frac{\partial}{\partial t'} - v\frac{\partial}{\partial z'}\right)f. \tag{2.3.6}$$

From the above derivation, the spatial and temporal derivatives prove to be transformed under the Lorentz transformation as

$$\begin{aligned}
&\frac{\partial}{\partial x} = \frac{\partial}{\partial x'}, \quad \frac{\partial}{\partial y} = \frac{\partial}{\partial y'}, \\
&\frac{\partial}{\partial z} = \gamma\left(\frac{\partial}{\partial z'} - \frac{v}{c^2}\frac{\partial}{\partial t'}\right), \\
&\frac{\partial}{\partial t} = \gamma\left(\frac{\partial}{\partial t'} - v\frac{\partial}{\partial z'}\right).
\end{aligned} \tag{2.3.7}$$

First, to find the transformation formulas for the field vectors $\boldsymbol{E}$ and $\boldsymbol{B}$, consider the transformation of the x component of the first equation of (2.3.1),

$$\frac{\partial E_z}{\partial y} - \frac{\partial E_y}{\partial z} + \frac{\partial B_x}{\partial t} = 0. \tag{2.3.8}$$

Applying the transformation (2.3.7) to (2.3.8) and rearranging the resultant equation properly, we get

$$\frac{\partial E_z}{\partial y'} - \frac{\partial}{\partial z'}\gamma\left(E_y + vB_x\right) + \frac{\partial}{\partial t'}\gamma\left(B_x + \frac{v}{c^2}E_y\right) = 0. \tag{2.3.9}$$

If we put

$$E_z' = E_z, \quad E_y' = \gamma\left(E_y + vB_x\right), \quad B_x' = \gamma\left(B_x + \frac{v}{c^2}E_y\right), \tag{2.3.10}$$

Eq. (2.3.9) becomes equal to the x component of the first equation of (2.3.3). Thus if Eqs. (2.3.8) is to be covariant under the Lorentz transformation, the y and z components of electric field and the x component of magnetic-flux density must be transformed as in (2.3.10). In a similar fashion, we can get the transformation formulas for the z and x components of electric field and the y component of magnetic-flux density by applying the transformation (2.3.7) to the y component of the first equation of (2.3.1). In addition, applying the transformation (2.3.7) to the z

component of the first equation of (2.3.1) and the second equation of (2.3.1), and combining to rearrange the resultant two equations, we get

$$\frac{\partial}{\partial x'}\gamma\left(E_y + vB_x\right) - \frac{\partial}{\partial y'}\gamma\left(E_x - vB_y\right) + \frac{\partial B_z}{\partial t'} = 0. \tag{2.3.11}$$

From the requirement that Eq. (2.3.11) agree with the z component of the first equation of (2.3.3), the transformation formulas for the x and y components of electric field and the z component of magnetic-flux density are obtained.

We can rearrange the transformation formulas for electric field intensity $\boldsymbol{E}$ and magnetic-flux density $\boldsymbol{B}$ obtained above in a vector form as follows:

$$\boldsymbol{E}'_{//} = \boldsymbol{E}_{//}, \quad \boldsymbol{E}'_{\perp} = \gamma\left(\boldsymbol{E} + \boldsymbol{v} \times \boldsymbol{B}\right)_{\perp}, \tag{2.3.12}$$

$$\boldsymbol{B}'_{//} = \boldsymbol{B}_{//}, \quad \boldsymbol{B}'_{\perp} = \gamma\left(\boldsymbol{B} - \frac{1}{c^2}\boldsymbol{v} \times \boldsymbol{E}\right)_{\perp}. \tag{2.3.13}$$

Similarly, applying the transformation (2.3.7) to (2.3.2), and requiring that the resultant equations agree with (2.3.4), we find the transformation formulas for electric-flux density $\boldsymbol{D}$ and magnetic field intensity $\boldsymbol{H}$, and those for current and charge densities, $\boldsymbol{J}$ and ρ in the following form:

$$\boldsymbol{D}'_{//} = \boldsymbol{D}_{//}, \quad \boldsymbol{D}'_{\perp} = \gamma\left(\boldsymbol{D} + \frac{1}{c^2}\boldsymbol{v} \times \boldsymbol{H}\right)_{\perp}, \tag{2,3,14}$$

$$\boldsymbol{H}'_{//} = \boldsymbol{H}_{//}, \quad \boldsymbol{H}'_{\perp} = \gamma\left(\boldsymbol{H} - \boldsymbol{v} \times \boldsymbol{D}\right)_{\perp}, \tag{2.3.15}$$

$$\boldsymbol{J}'_{//} = \gamma\left(\boldsymbol{J} - \rho\boldsymbol{v}\right)_{//}, \quad \boldsymbol{J}'_{\perp} = \boldsymbol{J}_{\perp}, \tag{2.3.16}$$

$$\rho' = \gamma\left(\rho - \frac{1}{c^2}\boldsymbol{v} \cdot \boldsymbol{J}\right). \tag{2.3.17}$$

Equations (2.3.12) ~ (2.3.17) express the transformation formulas for field vectors and those for charge and current densities from the inertial system I to the inertial system I'. The inverse transformation formulas from the system I' to the system I can be obtained by replacing the primed quantities by the unprimed ones, and $\boldsymbol{v}$ by $-\boldsymbol{v}$. In the above discussion, the transformation formulas for field vectors and those for charge and current densities were derived for a special case where the inertial system I' is uniformly moving in the z direction relative to the inertial system I. However, the transformation formulas (2.3.12) ~ (2.3.17) are represented in a vector form so that these formulas can be applied for the general case as well where the inertial system I' is uniformly moving in an arbitrary direction denoted by vector $\boldsymbol{v}$ relative to the inertial system I.

2.4 Constitutive Relations for Moving Media

According to the postulate of the constancy of the light velocity in inertial systems in the special theory of relativity, it is required that the velocity of light be equal to c in all inertial systems set up in vacuum. For this requirement to be met, the following constitutive relations must hold in all inertial systems in vacuum.

$$\boldsymbol{D} = \varepsilon_0 \boldsymbol{E}, \quad \boldsymbol{H} = \frac{1}{\mu_0}\boldsymbol{B}. \tag{2.4.1}$$

On the other hand, the constitutive relations for a material medium depend upon the velocity of it relative to an inertial system from which it is viewed. For a special case, let a homogeneous, isotropic and nondispersive medium with permittivity ε, permeability μ, and conductivity σ be moving with constant velocity $\boldsymbol{v}$ relative to the inertial system I. Then, let us try to find the constitutive relations for this medium as viewed from the inertial system I. To this end, the material medium is assumed to be at rest in the inertial system I', in which the following constitutive relations hold:

$$\boldsymbol{D}' = \varepsilon \boldsymbol{E}', \quad \boldsymbol{H}' = \frac{1}{\mu}\boldsymbol{B}', \quad \boldsymbol{J}_c' = \sigma \boldsymbol{E}'. \tag{2.4.2}$$

The corresponding constitutive relations in the inertial system I can be found by rewriting (2.4.2) in terms of the field vectors and current density in the inertial system I, with the aid of the transformation formulas for field vectors and current density obtained in the preceding section. For example, in order to find the constitutive relation in the system I corresponding to the first equation of (2.4.2), we first decompose the first equation of (2.4.2) into the component parallel to $\boldsymbol{v}$ and the component perpendicular to it, to each of which the transformation formulas (2.3.12) and (2.3.14) are applied. After that, the resultant equations are recombined. Thus the constitutive relation in the system I corresponding to the first constitutive relation of (2.4.2) can be obtained. In a similar fashion, the constitutive relation in the system I corresponding to the second equation of (2.4.2) can be found. The constitutive relations in the system I corresponding to the first and second equations of (2.4.2) obtained in this manner are expressed as follows:

$$\boldsymbol{D} + \frac{1}{c^2}\boldsymbol{v} \times \boldsymbol{H} = \varepsilon(\boldsymbol{E} + \boldsymbol{v} \times \boldsymbol{B}), \tag{2.4.3}$$

$$\boldsymbol{B} - \frac{1}{c^2}\boldsymbol{v} \times \boldsymbol{E} = \mu(\boldsymbol{H} - \boldsymbol{v} \times \boldsymbol{D}). \tag{2.4.4}$$

For the special case where $\boldsymbol{v}$ is in the z direction, the field vectors $\boldsymbol{D}$ and $\boldsymbol{B}$ are represented in terms of the field vectors $\boldsymbol{E}$ and $\boldsymbol{H}$ as

$$D = \varepsilon\hat{\alpha}\cdot E + \hat{\eta}\cdot H, \tag{2.4.5}$$

$$B = \mu\hat{\alpha}\cdot H - \hat{\eta}\cdot E, \tag{2.4.6}$$

where

$$\hat{\alpha} = \begin{bmatrix} 1 & 0 & 0 \\ 0 & \alpha & 0 \\ 0 & 0 & \alpha \end{bmatrix}, \quad \alpha = \frac{1-\beta^2}{1-n^2\beta^2},$$

$$\hat{\eta} = \begin{bmatrix} 0 & -\eta & 0 \\ \eta & 0 & 0 \\ 0 & 0 & 0 \end{bmatrix}, \quad \eta = \frac{(n^2-1)\beta}{(1-n^2\beta^2)c}, \quad n = \sqrt{\frac{\varepsilon\mu}{\varepsilon_0\mu_0}}. \tag{2.4.7}$$

For the case of vacuum, in particular, Eqs. (2.4.3) and (2.4.4) reduce to (2.4.1) as is expected. On the other hand, when viewed from an inertial system moving relative to a material medium, it behaves apparently like an anisotropic medium, and in the constitutive relations for it, the field vectors $\boldsymbol{D}$ and $\boldsymbol{B}$ depend on both the field vectors $\boldsymbol{E}$ and $\boldsymbol{H}$. This kind of medium is referred to as a bianisotropic medium [2.10].

Next, let us seek the constitutive relation in the system I corresponding to the third relation of (2.4.2). Consider only conduction current for the current in the system I'. Then, applying the transformations (2.3.16) and (2.3.12) to the third relation of (2.4.2), we get

$$J - \rho v = \gamma\sigma\left[E - \frac{1}{c^2}v(v\cdot E) + v\times B\right]. \tag{2.4.8}$$

The term $\rho\boldsymbol{v}$ on the left-hand side of (2.4.8) represents the convection current density produced by the motion of free charge with density ρ. Therefore, the total current density $\boldsymbol{J}$ minus the convection current density $\rho\boldsymbol{v}$ expresses the conduction current density. Hence, let $\boldsymbol{J}_c$ be the conduction current density in the system I, then Ohm's law in the system I reads as follows:

$$J_c = \gamma\sigma\left[E - \frac{1}{c^2}v(v\cdot E) + v\times B\right]. \tag{2.4.9}$$

The constitutive relations (2.4.3), (2.4.4), and (2.4.8) for moving media were first obtained by Minkowski in 1908 [2.6]. The Maxwell equations combined with the constitutive relations (2.4.3), (2.4.4), and (2.4.8) are often referred to as the Maxwell-Minkowski equations [2.8].

In the above discussion, we treated only the case where material media are moving uniformly relative to an inertial system. However, for the case where a material medium is moving nonuniformly or undergoing an accelerated motion such as a rotational motion relative to an inertial system, we could not find any inertial system in which the whole material medium stays at rest. In order to treat these cases in the same manner as for the case of uniformly moving media, we must define separate inertial systems at each point of the material medium at each time [2.9]. In other words, at each point of the material medium at each time, we must define an instantaneous rest frame, namely, an inertial system in which a particular point of the material medium is at rest at a particular instant of time. Then, if we can neglect the effects of strain due to the inertial forces such as the centrifugal and Coriolis forces appearing in a rotational motion, the constitutive relations (2.4.2) hold in the instantaneous rest frame as well. Thus, transforming the constitutive relations in the instantaneous rest frame to the inertial system I, we can find the constitutive relations at a particular point of the material medium at a particular instant of time as viewed from the system I. The constitutive relations obtained in this manner are identical in form to (2.4.3), (2.4.4), and (2.4.8). However, in these constitutive relations, we must use the value of the velocity of a material medium at a particular point at a particular instant of time.

2.5 Transformation of Frequency and Wave Numbers

In the preceding section, we have found the constitutive relations for a moving nondispersive medium by transforming the constitutive relations from the inertial system where the material medium is at rest to the inertial system where it is moving. However, if we wish to obtain the constitutive relations for moving dispersive media, the transformation formulas for frequency and wave numbers are required in addition to those for field vectors. In this section, we investigate how frequency and wave numbers are transformed under the Lorentz transformation. Consider the propagation of a uniform plane wave in a homogeneous medium. Let $\boldsymbol{E}$ and $\boldsymbol{E}'$ be the electric field vectors in the systems I and I'. Then they are expressed in complex form as

$$\boldsymbol{E} = \mathrm{Re}\left[\tilde{\boldsymbol{E}} e^{j(\omega t - \boldsymbol{k}\cdot\boldsymbol{r})}\right],$$

$$\boldsymbol{E}' = \mathrm{Re}\left[\tilde{\boldsymbol{E}}' e^{j(\omega' t' - \boldsymbol{k}'\cdot\boldsymbol{r}')}\right], \tag{2.5.1}$$

where $\tilde{\boldsymbol{E}}$ and $\tilde{\boldsymbol{E}}'$ are complex amplitudes, $\boldsymbol{r}$ and $\boldsymbol{r}'$ position vectors, ω and ω' angular frequencies, and $\boldsymbol{k}$ and $\boldsymbol{k}'$ wave vectors. The Cartesian components of $\boldsymbol{k}$ and $\boldsymbol{k}'$ are denoted by (k_x, k_y, k_z) and (k'_x, k'_y, k'_z), respectively.

Now, let us note the phase of a plane wave at a particular position at a particular time in the system I. Then the value of the phase is kept invariant if viewed from the system I'. In other words, the phase is an invariant under the Lorentz transformation. The invariance of the phase is expressed as

$$\omega t - \boldsymbol{k}\cdot\boldsymbol{r} = \omega' t' - \boldsymbol{k}'\cdot\boldsymbol{r}'. \tag{2.5.2}$$

Rewriting the above equation in Cartesian components leads to

$$\omega t-\left(k_x x+k_y y+k_z z\right)=\omega' t'-\left(k'_x x'+k'_y y'+k'_z z'\right). \tag{2.5.3}$$

Rearranging after inserting (2.2.9) in the left-hand side of (2.5.3), we get

$$\omega t-\left(k_x x+k_y y+k_z z\right)$$
$$=\gamma\left(\omega-vk_z\right)t'-k_x x'-k_y y'-\gamma\left(k_z-\frac{v}{c^2}\omega\right)z'. \tag{2.5.4}$$

Comparing the right-hand sides of (2.5.3) and (2.5.4) to each other, we find for the transformation formulas for frequency and wave numbers,

$$k'_x=k_x,\quad k'_y=k_y,\quad k'_z=\gamma\left(k_z-\frac{v}{c^2}\omega\right),\quad \omega'=\gamma\left(\omega-vk_z\right). \tag{2.5.5}$$

Referring to (2.2.8), we recognize that the components of the wave vector $\boldsymbol{k}$ are transformed like the spatial coordinates and ω/c^2 like the temporal coordinate. Therefore, for the case where the system I' is moving in an arbitrary direction relative to the system I, the transformation formulas for frequency and wave numbers are readily obtained by referring to (2.2.10) as

$$\boldsymbol{k}'_\perp=\boldsymbol{k}_\perp,\quad \boldsymbol{k}'_{//}=\gamma\left(\boldsymbol{k}_{//}-\boldsymbol{v}\frac{\omega}{c^2}\right),\quad \omega'=\gamma\left(\omega-\boldsymbol{v}\cdot\boldsymbol{k}\right). \tag{2.5.6}$$

From the transformation formulas for wave numbers in (2.5.5) or (2.5.6), we can find how the wavelength and the direction of propagation of a plane wave change when we move from one inertial frame to another. The last equations of (2.5.5) and (2.5.6) denote the change in frequency or the Doppler shift in frequency produced by the relative motion of two inertial systems. As is evident from the above discussion, in relativity, the frequency and wave numbers are transformed as a cluster, or more precisely, as a four-vector in the same manner as the space-time coordinates.

With the aid of the transformation formula for frequency, we can find the constitutive relations for a moving temporally dispersive medium or a moving medium with frequency dispersion. Consider an electromagnetic field varying sinusoidally in time at a single frequency. Then, as shown in (1.6.9), the constitutive relations in the inertial system I', where a medium with frequency dispersion is at rest, are given by

$$\tilde{\boldsymbol{D}}'=\hat{\varepsilon}(\omega')\cdot\tilde{\boldsymbol{E}}',\quad \tilde{\boldsymbol{B}}'=\hat{\mu}(\omega')\cdot\tilde{\boldsymbol{H}}'. \tag{2.5.7}$$

Transforming (2.5.7) to the inertial system I, as we did in obtaining (2.4.3) and (2.4.4), we find the following constitutive relations for a dispersive medium moving with constant velocity $\boldsymbol{v}$ relative to the system I:

$$\tilde{\boldsymbol{D}} + \frac{1}{c^2}\boldsymbol{v} \times \tilde{\boldsymbol{H}} = \hat{\varepsilon}(\omega - \boldsymbol{v}\cdot\boldsymbol{k})\cdot\left(\tilde{\boldsymbol{E}} + \boldsymbol{v} \times \tilde{\boldsymbol{B}}\right), \tag{2.5.8}$$

$$\tilde{\boldsymbol{B}} - \frac{1}{c^2}\boldsymbol{v} \times \tilde{\boldsymbol{E}} = \hat{\mu}(\omega - \boldsymbol{v}\cdot\boldsymbol{k})\cdot\left(\tilde{\boldsymbol{H}} - \boldsymbol{v} \times \tilde{\boldsymbol{D}}\right), \tag{2.5.9}$$

where $\hat{\varepsilon}(\omega - \boldsymbol{v}\cdot\boldsymbol{k})$ and $\hat{\mu}(\omega - \boldsymbol{v}\cdot\boldsymbol{k})$ denote $\hat{\varepsilon}(\omega')$ and $\hat{\mu}(\omega')$ with ω' replaced by the frequency and wave numbers in the system I with the aid of the last equation of (2.5.6). As seen from (2.5.8) and (2.5.9), if a medium only with frequency dispersion is viewed from an inertial system moving relative to the rest frame of the medium, not only the frequency dispersion but also the wave number dispersion appear in that system.

As an example, let us consider cold isotropic electron plasma for which the effect of the thermal motion of electrons is negligible. The equivalent permittivity tensor in the rest frame of the electron plasma, $\hat{\varepsilon}(\omega')$ is given by

$$\hat{\varepsilon}(\omega') = \hat{u}\varepsilon_p = \hat{u}\varepsilon_0\left[1 - \left(\frac{\omega_p}{\omega'}\right)^2\right], \tag{2.5.10}$$

where $\hat{u}$ denotes a unit tensor and ω_p the plasma angular frequency for the electron plasma. The permeability of the electron plasma is assumed to be equal to μ_0. In the inertial system I in which the electron plasma is moving with constant velocity $\boldsymbol{v}$, we can rewrite $\hat{\varepsilon}(\omega')$ in terms of the frequency and wave numbers in the system I, with the aid of the last equation of (2.5.6), as

$$\hat{\varepsilon}(\omega') = \hat{u}\varepsilon_0\left[1 - \frac{\omega_p^2}{\gamma^2(\omega - \boldsymbol{v}\cdot\boldsymbol{k})^2}\right] = \hat{\varepsilon}(\omega - \boldsymbol{v}\cdot\boldsymbol{k}). \tag{2.5.11}$$

The constitutive relations for an isotropic electron plasma moving with constant velocity $\boldsymbol{v}$ can be found by inserting (2.5.11) in (2.5.8), and $\hat{\mu}(\omega - v\cdot k) = \hat{u}\mu_0$ in (2.5.9).

2.6 Integral Representation of the Maxwell Equations in Moving Systems

Consider a closed contour C moving with constant velocity $\boldsymbol{v}$ relative to the inertial system I, and an open surface S bounded by the closed contour C, as shown in Fig. 2.2. Then, let us discuss the integral representation of the Maxwell equations for the closed contour C and the open surface S. First, we consider Faraday's induction law, which states that if the amount of the magnetic flux linking the closed contour C is temporally changed, the electromotive force is produced around C with the magnitude equal to the negative time rate of change of the magnetic flux linking the contour C. In general, the closed contour may be at rest or moving relative to the observer, or it may be deformed in an arbitrary manner. For a closed contour at rest, Faraday's induction law is given by (1.1.1) or (1.2.4). On the other hand, for the case where a closed contour is moving with velocity $\boldsymbol{v}$, it reads as

$$\oint_C (\boldsymbol{E} + \boldsymbol{v} \times \boldsymbol{B}) \cdot d\boldsymbol{l} = -\frac{d}{dt}\int_S \boldsymbol{B} \cdot d\boldsymbol{s}. \tag{2.6.1}$$

The reason why the left-hand side of (2.6.1) expresses the electromotive force induced along a closed contour C moving with velocity $\boldsymbol{v}$ is understood in the following manner. First note that the electromotive force induced along a closed contour C is equal to the work done by an electromagnetic field when a unit electric charge is carried around the closed contour C by the electromagnetic force. The integrand $\boldsymbol{E} + \boldsymbol{v} \times \boldsymbol{B}$ in the line integral on the left-hand side of (2.6.1) corresponds to the force exerted on a unit charge by an electromagnetic field at a point on the closed contour C moving with velocity $\boldsymbol{v}$. Hence the line integral on the left-hand side of (2.6.1) is equal to the total work done on a unit charge by the field when it is carried around the closed contour C. This work is, by definition, the electromotive force induced along the closed contour C. In addition, it should be noted that the temporal change of the magnetic flux linking the contour C in (2.6.1)

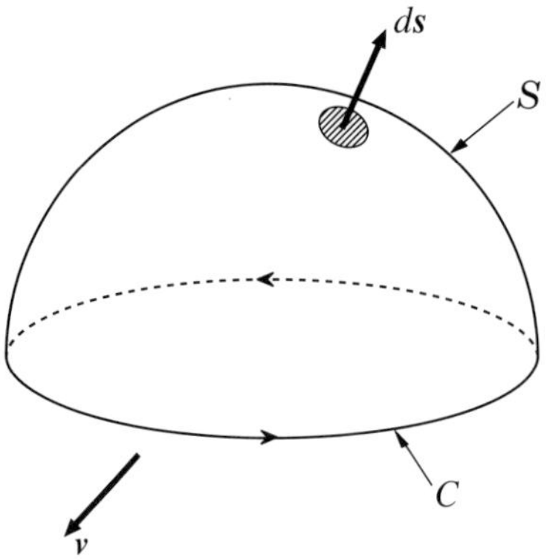

Fig. 2.2. A moving closed contour C and an open surface S bounded by C

includes not only the temporal change of $\boldsymbol{B}$ itself but also the change of the magnetic flux due to the motion of the contour C. On the other hand, for the case of (1.1.1) where the contour C is at rest, only the temporal change of $\boldsymbol{B}$ itself contributes to the induction of the electromotive force around the contour C.

Next, we discuss Ampere-Maxwell's law. According to this law, if current flows through any closed contour C or the amount of the electric flux linking the closed contour C is temporally changed, the magnetic field is produced around the closed contour C. Specifying Ampere-Maxwell's law for a closed contour C moving with constant velocity $\boldsymbol{v}$, and an open surface S bounded by the closed contour C, we get

$$\oint_C (\boldsymbol{H} - \boldsymbol{v} \times \boldsymbol{D}) \cdot d\boldsymbol{l} = \frac{d}{dt}\int_S \boldsymbol{D} \cdot d\boldsymbol{s} + \int_S (\boldsymbol{J} - \rho \boldsymbol{v}) \cdot d\boldsymbol{s}. \tag{2.6.2}$$

By analogy with the left-hand side of (2.6.1), we can regard the left-hand side of (2.6.2) as the magnetomotive force induced around a moving closed contour C. In addition, the term $-\rho \boldsymbol{v}$ appears in the integrand of the second surface integral on the right-hand side of (2.6.2). This is because the current of density $-\rho \boldsymbol{v}$ flows through the surface S when it passes through the charge distribution of density ρ. Equations (2.6.1) and (2.6.2) are Faraday's induction law and Ampere-Maxwell's law specified for a moving closed contour C and an open surface S bounded by the closed contour C. Although Eqs. (2.6.1) and (2.6.2) apparently appear different from (1.2.4) and (1.2.5), the physical meanings of the former are essentially identical to those of the latter.

From the above discussion, we have found that Faraday's induction law and Ampere-Maxwell's law should be reformulated for a moving closed contour C and an open surface S bounded by the closed contour C, as shown in (2.6.1) and (2.6.2). On the other hand, Gauss' laws for the electric and magnetic fields are expressed mathematically in the same form whether any closed surface S for describing the laws is at rest or moving. Hence, for any closed surface S moving with velocity $\boldsymbol{v}$ relative to the inertial system I, Gauss' laws for the electric and magnetic fields are represented, respectively, as

$$\oint_S \boldsymbol{D} \cdot d\boldsymbol{s} = \int_V \rho dv, \tag{2.6.3}$$

$$\oint_S \boldsymbol{B} \cdot d\boldsymbol{s} = 0, \tag{2.6.4}$$

where the surface integral and the volume integral are carried out over any closed surface S moving with velocity $\boldsymbol{v}$ and a volume V enclosed by S.

2.7 Boundary Conditions for a Moving Boundary

As shown in Fig. 2.3, let the boundary S between regions I and II be moving in an arbitrary direction with constant velocity $\boldsymbol{v}$ relative to the inertial system I, and then let us discuss the boundary conditions to be imposed on field vectors at the moving boundary S. We first consider, as in Fig. 1.4, a small thin cylindrical domain around a point P on the boundary S, which is moving together with the boundary. To this small cylindrical domain, we apply Gauss' laws for the electric and magnetic fields, (2.6.3) and (2.6.4), as in Sect. 1.4. Then, if we shrink the small cylindrical domain around the point P, keeping the thickness of it much less than the diameter of the top and bottom surfaces of it, we get the following boundary conditions for the normal components of electric-flux density $\boldsymbol{D}$ and magnetic-flux density $\boldsymbol{B}$ to satisfy at the boundary:

$$\boldsymbol{n}\cdot\left(\boldsymbol{D}_1 - \boldsymbol{D}_2\right) = \xi, \tag{2.7.1}$$

$$\boldsymbol{n}\cdot\left(\boldsymbol{B}_1 - \boldsymbol{B}_2\right) = 0, \tag{2.7.2}$$

where $\boldsymbol{n}$ is a unit vector normal to the boundary directed from region II to region I, and ξ denotes the surface charge density on the boundary, with subscripts 1 and 2 referring to field vectors in regions I and II. Comparing (2.7.1) and (2.7.2) with (1.4.1) and (1.4.2), we recognize that the motion of the boundary does not affect the boundary conditions for the components of field vectors $\boldsymbol{D}$ and $\boldsymbol{B}$ normal to the boundary.

Next, as shown in Fig. 1.5, we consider a small rectangular contour C normal to the boundary surface S and intersecting S, which is moving together with the boundary. To this small rectangular contour C, we apply (2.6.1) and (2.6.2). Then, as in **1.4**, we shrink the small rectangular contour C around the point P on the boundary surface S, keeping the length of the sides of the rectangular contour normal to the boundary much less than the length of those parallel to the boundary. In this manner, we find the following boundary conditions for the components of field vectors $\boldsymbol{E}$ and $\boldsymbol{H}$ tangential to the boundary surface to satisfy at the boundary:

$$\boldsymbol{n}\times\left[\left(\boldsymbol{E}_1 + \boldsymbol{v}\times\boldsymbol{B}_1\right) - \left(\boldsymbol{E}_2 + \boldsymbol{v}\times\boldsymbol{B}_2\right)\right] = 0, \tag{2.7.3}$$

$$\boldsymbol{n}\times\left[\left(\boldsymbol{H}_1 - \boldsymbol{v}\times\boldsymbol{D}_1\right) - \left(\boldsymbol{H}_2 - \boldsymbol{v}\times\boldsymbol{D}_2\right)\right] = \boldsymbol{K} - \xi\boldsymbol{v}, \tag{2.7.4}$$

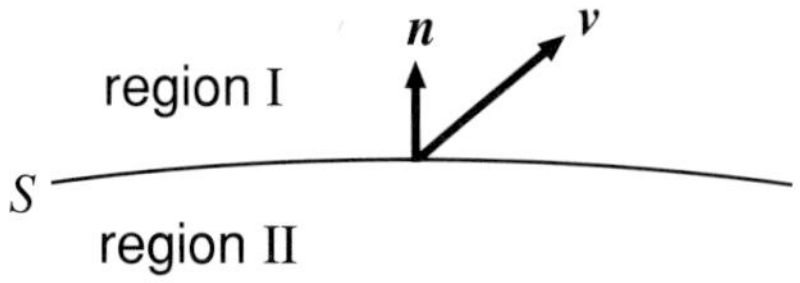

Fig. 2.3. A moving boundary S

$\boldsymbol{K}$ denoting the surface current density on the boundary surface. Expanding the vector triple products in the left-hand sides of (2.7.3) and (2.7.4), and using the relations (2.7.1) and (2.7.2) in the resultant equations, we get the following relations:

$$\boldsymbol{n} \times (\boldsymbol{E}_1 - \boldsymbol{E}_2) - (\boldsymbol{n} \cdot \boldsymbol{v})(\boldsymbol{B}_1 - \boldsymbol{B}_2) = 0, \tag{2.7.5}$$

$$\boldsymbol{n} \times (\boldsymbol{H}_1 - \boldsymbol{H}_2) + (\boldsymbol{n} \cdot \boldsymbol{v})(\boldsymbol{D}_1 - \boldsymbol{D}_2) = \boldsymbol{K}. \tag{2.7.6}$$

Note that the surface current density $\boldsymbol{K}$ includes not only the component parallel to the boundary surface but also the component normal to it, which is produced by the motion of the surface charge of density ξ. In fact, we find, from (2.7.6) and (2.7.1), the relation

$$\boldsymbol{n} \cdot \boldsymbol{K} = \boldsymbol{n} \cdot (\xi \boldsymbol{v}), \tag{2.7.7}$$

which proves the above statement concerning the surface current density $\boldsymbol{K}$.

It should also be noted that the boundary conditions (2.7.5) and (2.7.6) depend only upon the velocity component of the boundary normal to it, $\boldsymbol{n} \cdot \boldsymbol{v}$ but not upon the parallel velocity component of it. Hence, for the case where the boundary is moving parallel to itself, or for the case of $\boldsymbol{n} \cdot \boldsymbol{v} = 0$, the boundary conditions (2.7.5) and (2.7.6) reduce to

$$\boldsymbol{n} \times (\boldsymbol{E}_1 - \boldsymbol{E}_2) = 0, \tag{2.7.8}$$

$$\boldsymbol{n} \times (\boldsymbol{H}_1 - \boldsymbol{H}_2) = \boldsymbol{K}, \tag{2.7.9}$$

which are identical to the boundary conditions for a stationary boundary, (1.4.7) and (1.4.8). The boundary conditions (2.7.8) and (2.7.9) are very useful from the standpoint of practical applications.

Finally, we supplement a note on the motion of the boundary surface. It is that the motion of the boundary surface S in Fig. 2.3 does not always coincide with the motion of the material media occupying regions I and II. For example, let us consider the case where region I is vacuum and region II is occupied by a dielectric,

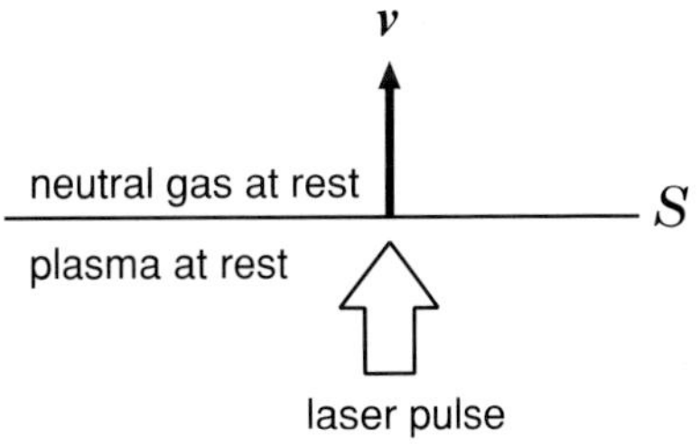

Fig. 2.4. Neutral gas and plasma at rest separated by a moving boundary

which is moving in an arbitrary direction with velocity $\boldsymbol{v}$. For this case, it is true that the motion of the boundary coincides with the motion of the medium. On the other hand, as shown in Fig. 2.4, let an intense laser pulse be propagating through a neutral gas, which is ionized by it. For this case, the front of the laser pulse constitutes a moving boundary, which leaves an ionized gas (plasma) backward and a neutral gas forward. Then, note that both the neutral and ionized gases remain stationary while the boundary between them is moving with the laser pulse. Needless to say, the boundary conditions at a moving boundary discussed in this section can be applied to either of the cases exemplified above.

2.8 Equivalence of Energy and Mass

Let us consider the equivalence of energy and the corresponding inertial mass in the special theory of relativity. In order to make our discussion comprehensible to the audience, we introduce a thought experiment that was originally proposed by Einstein [2.13, 2.14]. In the thought experiment (see Fig. 2.5), we consider an isolated hollow cylindrical box with mass M and length l, which is at rest at the initial state. From one end wall of the cylindrical box, a narrow electromagnetic wave packet, say a light pulse, is emitted, and this wave packet is absorbed into the other end wall after propagating in vacuum by the distance l. In this process, we observe the exchange of energy and momentum between the box and the electromagnetic wave packet. As will be clarified in the subsequent section, an electromagnetic field carries the corresponding momentum, the density of which is given by $\boldsymbol{S}/c^2$, $\boldsymbol{S}$ being the Poynting vector for the electromagnetic field. Let the wave packet be propagating in the z direction. Then it carries the total momentum equal to

$$G = \int_V \frac{\boldsymbol{S}}{c^2} \cdot \boldsymbol{i}_z dv = \int_V \frac{S}{c^2} dv , \tag{2.8.1}$$

where the integration is carried out over the volume the wave packet occupies, and the spread of the wave packet in the z direction is assumed to be much less than l. On the other hand, as shown in (1.8.22) for a plane wave, we have the following relation between the Poynting vector $\boldsymbol{S}$ and the energy density w for an electromagnetic wave propagating in the z direction in vacuum:

$$\frac{\boldsymbol{S}}{w} = \boldsymbol{i}_z c . \tag{2.8.2}$$

The total energy carried by the wave packet, W is expressed in terms of the total momentum G as follows:

$$W = \int_V w dv = \int_V \frac{S}{c} dv = cG . \tag{2.8.3}$$

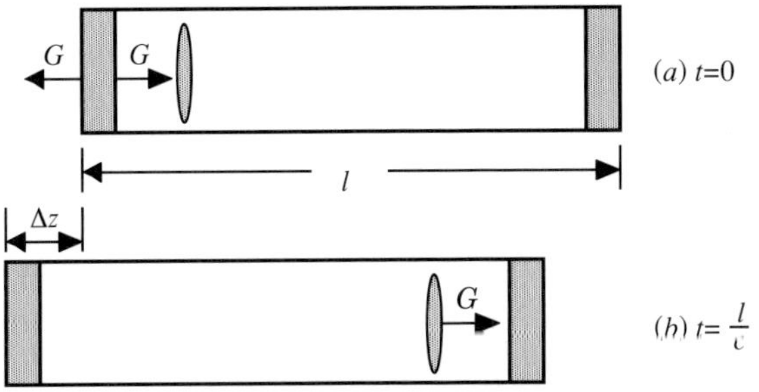

Fig. 2.5. A thought experiment called Einstein's box

We are now ready to discuss the exchange of energy and momentum between the cylindrical box and the wave packet. At $t = 0$, let an electromagnetic wave packet with energy W and momentum G be emitted from the left end wall of the hollow cylindrical box. Then, the cylindrical box is kicked in the negative z direction by recoil from the wave packet, and it gets momentum $-G$, beginning to move in that direction with constant velocity v. The velocity v acquired by the box is found from the relation

$$v = \frac{G}{M} = \frac{W}{cM}, \tag{2.8.4}$$

where $v << c$ is assumed, and the momentum-velocity relation in Newtonian mechanics is used for the motion of the cylindrical box.

The box continues to move until the wave packet impinges upon the right end wall of the box at $t = l/c$, when the wave packet gives momentum G back to the box, which is in the positive z direction. This time, the momentum G makes the box move in the positive z direction with velocity v. Hence the net effect of the wave packet on the box is to shift the center of mass of the box in the negative z direction by the amount,

$$\Delta z = v \times \frac{l}{c} = \frac{Wl}{c^2 M}. \tag{2.8.5}$$

Although the box has shifted in the negative z direction by Δz, the center of mass of the whole system including the box and the wave packet must remain at rest, since the whole system is isolated. Hence let us assume that the wave packet has carried away, from the box, the mass corresponding to the energy it carries by m. Then, if the center of mass for the whole system is to remain fixed, we must have

$$m(l - \Delta z) - (M - m)\Delta z = ml - M\Delta z = 0. \tag{2.8.6}$$

Substituting (2.8.5) in (2.8.6), we get the relation

$$W = mc^2 . \tag{2.8.7}$$

The relation (2.8.7), which represents the equivalence of energy and mass, is one of the highlights in Einstein's relativity, having made a profound impact on modern physics. Although we considered electromagnetic energy in deriving (2.8.7), it is a universal law that is applicable to all types of energy, say nuclear, thermal or mechanical. In addition, note that the law has been well confirmed by experiments.

2.9 Relativistic Mechanics for a Material Particle

Let us discuss how Newtonian mechanics for the motion of a material particle under the influence of external forces is modified in the special theory of relativity. For this purpose, we assume that the equation of motion and the energy conservation relation for a material particle in relativistic mechanics can be written formally in the same manner as in Newtonian mechanics. Namely, for the equation of motion for a material particle, we have

$$\frac{d\boldsymbol{G}}{dt} = \boldsymbol{F} , \tag{2.9.1}$$

where $\boldsymbol{F}$ is an external force, and $\boldsymbol{G}$ denotes the momentum of a material particle, being expressed as

$$\boldsymbol{G} = m\boldsymbol{v} , \tag{2.9.2}$$

m and $\boldsymbol{v}$ being the relativistic mass and the velocity of a material particle, respectively. Although the momentum of a material particle in relativistic mechanics is formally expressed in the same way as in Newtonian mechanics, as shown in (2.9.2), it should be noted that the mass of a material particle in the former is allowed to depend upon the magnitude of its velocity while it is constant independent of the velocity in the latter.

In addition, for the energy conservation relation for a material particle under the influence of external force $\boldsymbol{F}$, we have

$$\frac{dW}{dt} = \boldsymbol{F} \cdot \boldsymbol{v} , \tag{2.9.3}$$

where W is the energy carried by a material particle, and the right-hand side represents the work done per second on the particle by the external force $\boldsymbol{F}$.

From (2.9.3) and (2.9.1), we find

$$dW = \boldsymbol{F} \cdot d\boldsymbol{l} = \boldsymbol{F} \cdot \boldsymbol{v} dt = \boldsymbol{v} \cdot d\boldsymbol{G} , \tag{2.9.4}$$

where $d\boldsymbol{l}$ denotes a differential vector line element. Multiplying m on both sides of (2.9.4), and using the relation (2.8.7) in the left-hand side, we get

$$\frac{W}{c^2}dW = \boldsymbol{G}\cdot d\boldsymbol{G}. \tag{2.9.5}$$

Let a material particle be at rest at the initial state, and accelerated to the velocity $\boldsymbol{v}$. Then, from (2.9.5), we obtain

$$W^2 = c^2G^2 + W_0^2, \tag{2.9.6}$$

where W and G are, respectively, the energy and the magnitude of momentum corresponding to the velocity $\boldsymbol{v}$ of a material particle, and W_0 the energy of a material particle at rest.

With the aid of (2.8.7) and (2.9.2), we can rewrite (2.9.6) as

$$W = \frac{W_0}{\sqrt{1-\left(\frac{v}{c}\right)^2}}. \tag{2.9.7}$$

For $v << c$, Eq. (2.9.7) can be expanded as

$$W = W_0 + \frac{1}{2}\frac{W_0}{c^2}v^2. \tag{2.9.8}$$

If Eq. (2.9.8) should be in accordance with Newtonian mechanics, the second term on the right-hand side must correspond to the kinetic energy of a particle. Hence we get the relation

$$\frac{W_0}{c^2} = m_0, \tag{2.9.9}$$

m_0 being the inertial mass of a material particle in Newtonian mechanics. In relativistic mechanics, m_0 and W_0 are referred to as the rest mass and rest energy of a material particle, respectively. Inserting the relations (2.8.7) and (2.9.9) in (2.9.7), we obtain

$$m = \frac{m_0}{\sqrt{1-\left(\frac{v}{c}\right)^2}}, \tag{2.9.10}$$

m representing the relativistic mass of a material particle, which corresponds to the sum of the rest energy and kinetic energy of it.

2.10 Relativistic Equation of Motion for a Moving Charged Particle

We consider the relativistic motion of a charged particle in electromagnetic fields. Let a charged particle with rest mass m_0 and charge q be moving with an arbitrary velocity $\boldsymbol{v}$. Then this charged particle has momentum $\boldsymbol{G}$ and kinetic energy K defined by

$$G = mv = \frac{m_0 v}{\sqrt{1-\left(\frac{v}{c}\right)^2}} = \gamma m_0 v, \tag{2.10.1}$$

$$K = \frac{m_0 c^2}{\sqrt{1-\left(\frac{v}{c}\right)^2}} - m_0 c^2 = (\gamma - 1) m_0 c^2, \tag{2.10.2}$$

with

$$\gamma = \frac{1}{\sqrt{1-\left(\frac{v}{c}\right)^2}}. \tag{2.10.3}$$

With the aid of (2.9.1) and (2.10.1), the relativistic equation of motion for a charged particle moving in electric field $\boldsymbol{E}$ and magnetic-flux density $\boldsymbol{B}$ takes the form

$$\frac{d\boldsymbol{G}}{dt} = \frac{d}{dt}(\gamma m_0 \boldsymbol{v}) = q(\boldsymbol{E} + \boldsymbol{v} \times \boldsymbol{B}), \tag{2.10.4}$$

where the right-hand side expresses the Lorentz force exerted on the charged particle.

The electric field acting upon the charged particle does work on it. The amount of work done per second on the charged particle by the electric field is given by $q\boldsymbol{v} \cdot \boldsymbol{E}$. This work changes the kinetic energy of the particle. Hence the following energy conservation relation holds for the charged particle:

$$\frac{dK}{dt} = \frac{d}{dt}\left(\gamma m_0 c^2\right) = q\boldsymbol{v} \cdot \boldsymbol{E}. \tag{2.10.5}$$

The kinetic energy of the charged particle increases if $q\boldsymbol{v} \cdot \boldsymbol{E} > 0$, while it decreases if $q\boldsymbol{v} \cdot \boldsymbol{E} < 0$. Thus the electric field does work on the charged particle, whereas the magnetic field does not because the force due to it is perpendicular to the velocity of the particle. With the aid of (2.10.5), the equation of motion (2.10.4) can be rewritten as

$$\frac{d\boldsymbol{v}}{dt} = \frac{q}{\gamma m_0}\left[\boldsymbol{E} - \frac{\boldsymbol{v}}{c^2}(\boldsymbol{v} \cdot \boldsymbol{E}) + \boldsymbol{v} \times \boldsymbol{B}\right]. \tag{2.10.6}$$

For a special case, let us discuss the energy conservation relation for a charged particle moving in an electrostatic field. Since an electrostatic field can be expressed in terms of a scalar potential ϕ as $\boldsymbol{E} = -\nabla\phi$, Eq. (2.10.5) reduces to

$$\frac{d}{dt}\left(\gamma m_0 c^2\right) = -q\boldsymbol{v}\cdot\nabla\phi. \tag{2.10.7}$$

Rewriting the right-hand side of (2.10.7), we get

$$\frac{d}{dt}\left(\gamma m_0 c^2\right) = -q\left(\frac{\partial\phi}{\partial x}\frac{dx}{dt} + \frac{\partial\phi}{\partial y}\frac{dy}{dt} + \frac{\partial\phi}{\partial z}\frac{dz}{dt}\right) = -q\frac{d\phi}{dt}, \tag{2.10.8}$$

or

$$\frac{d}{dt}\left[\gamma m_0 c^2 + q\phi\right] = 0. \tag{2.10.9}$$

Integrating (2.10.9) with respect to t, we obtain the following energy conservation relation for a charged particle in an electrostatic field:

$$\gamma m_0 c^2 + q\phi = C, \tag{2.10.10}$$

C being a constant of integration. For example, if $v = 0$ when $\phi = 0$, we have $C = m_0c^2$.

2.11 Energy and Momentum Conservation Laws for a System of Charged Particles and Electromagnetic Fields

We discuss the energy and momentum conservation relations for a system composed of charged particles and electromagnetic fields in vacuum. To this end, let us consider the interaction of N charged particles and electromagnetic fields distributed in a volume V enclosed by a closed surface S, taking the relativistic effect into account. The charge density ρ_i and the current density $\boldsymbol{J}_i$ due to the ith charged particle located at $\boldsymbol{r}_i$ are expressed in terms of the Dirac delta function as

$$\rho_i(\boldsymbol{r},t) = q_i\delta\left[\boldsymbol{r} - \boldsymbol{r}_i(t)\right], \tag{2.11.1}$$

$$\boldsymbol{J}_i(\boldsymbol{r},t) = q_i\boldsymbol{v}_i\delta\left[\boldsymbol{r} - \boldsymbol{r}_i(t)\right], \tag{2.11.2}$$

where q_i and $\boldsymbol{v}_i$ are the charge and velocity of the ith charged particle. Summing up all the contributions from N charged particles, we have for the total charge and current densities,

$$\rho(\boldsymbol{r},t) = \sum_{i=1}^{N}\rho_i(\boldsymbol{r}) = \sum_{i=1}^{N} q_i\delta\left[\boldsymbol{r} - \boldsymbol{r}_i(t)\right], \tag{2.11.3}$$

$$\boldsymbol{J}(\boldsymbol{r},t) = \sum_{i=1}^{N}\boldsymbol{J}_i(\boldsymbol{r},t) = \sum_{i=1}^{N} q_i\boldsymbol{v}_i\delta\left[\boldsymbol{r} - \boldsymbol{r}_i(t)\right]. \tag{2.11.4}$$

First, let us consider the energy conservation law for a system composed of charged particles and electromagnetic fields. With the aid of (2.11.2), we get for the mechanical energy of the ith charged particle,

$$\frac{d}{dt}\left(\frac{m_{i0}c^2}{\sqrt{1-\left(\frac{v_i}{c}\right)^2}}\right) = \int_V q_i \boldsymbol{v}_i \delta(\boldsymbol{r}-\boldsymbol{r}_i)\cdot \boldsymbol{E}(\boldsymbol{r},t)dv$$

$$= \int_V \boldsymbol{J}_i(\boldsymbol{r},t)\cdot \boldsymbol{E}(\boldsymbol{r},t)dv. \qquad (2.11.5)$$

m_{i0} being the rest mass of the ith charged particle. From (2.11.4) and (2.11.5), the total mechanical energy of N charged particles, $W_m(t)$ is found to satisfy

$$\frac{dW_m(t)}{dt} = \int_V \boldsymbol{J}(\boldsymbol{r},t)\cdot \boldsymbol{E}(\boldsymbol{r},t)dv, \qquad (2.11.6)$$

where

$$W_m(t) = \sum_{i=1}^{N} \frac{m_{i0}c^2}{\sqrt{1-\left(\frac{v_i}{c}\right)^2}}. \qquad (2.11.7)$$

With the aid of the Maxwell equations in vacuum, (1.1.13) ~ (1.1.16), the right-hand side of (2.11.16) can be rewritten as

$$\frac{dW_m}{dt} = \int_V \left[\boldsymbol{E}\cdot\left(\nabla\times\frac{\boldsymbol{B}}{\mu_0} - \frac{\partial}{\partial t}\varepsilon_0\boldsymbol{E}\right) - \frac{\boldsymbol{B}}{\mu_0}\cdot\left(\nabla\times\boldsymbol{E}+\frac{\partial\boldsymbol{B}}{\partial t}\right)\right]dv. \qquad (2.11.8)$$

From (2.11.8), we immediately get the energy conservation law for a system composed of charged particles and electromagnetic fields,

$$\frac{d}{dt}\left(W_m + W_f\right) = -\oint_S \boldsymbol{S}\cdot\boldsymbol{n}ds, \qquad (2.11.9)$$

where $\boldsymbol{n}$ is a unit vector outwardly normal to a closed surface S, and W_f and $\boldsymbol{S}$ are, respectively, the total electromagnetic energy stored in the volume V and the electromagnetic power flow density, being expressed as

$$W_f = \frac{1}{2}\int_V \left(\varepsilon_0|\boldsymbol{E}|^2 + \frac{1}{\mu_0}|\boldsymbol{B}|^2\right)dv = \frac{1}{2}\int_V \left(\varepsilon_0|\boldsymbol{E}|^2 + \mu_0|\boldsymbol{H}|^2\right)dv, \qquad (2.11.10)$$

$$\boldsymbol{S} = \boldsymbol{E}\times\frac{\boldsymbol{B}}{\mu_0} = \boldsymbol{E}\times\boldsymbol{H}. \qquad (2.11.11)$$

The law of energy conservation, (2.11.9) for a system composed of charged particles and electromagnetic fields represents that the temporal increment of the total mechanical and field energy is supplied by the electromagnetic power flow from the outside of the closed surface S. If the latter vanishes, the sum of the mechanical energy and the field energy stored in the volume V is conserved.

Let us now discuss the law of momentum conservation for a system composed of charged particles and electromagnetic fields. As shown in (2.10.4), the relativistic equation of motion for a charged particle at the position $\boldsymbol{r}_i$ is given by

$$\frac{d\boldsymbol{G}_i(t)}{dt} = q_i\left[\boldsymbol{E}(\boldsymbol{r}_i,t) + \boldsymbol{v}_i \times \boldsymbol{B}(\boldsymbol{r}_i,t)\right], \quad (i = 1,2,\cdots N), \tag{2.11.12}$$

with

$$\boldsymbol{G}_i(t) = \frac{m_{i0}\boldsymbol{v}}{\sqrt{1-\left(\frac{v_i}{c}\right)^2}}, \tag{2.11.13}$$

where $\boldsymbol{G}_i$ is the mechanical momentum of the ith particle.

With the aid of (2.11.12) and (2.11.1) ~ (2.11.4), the total mechanical momentum of N charged particles, $\boldsymbol{G}_m$ is found to satisfy

$$\begin{aligned}\frac{d\boldsymbol{G}_m(t)}{dt} &= \sum_{i=1}^{N}\frac{d\boldsymbol{G}_i(t)}{dt}\\ &= \sum_{i=1}^{N}\int_V\left[q_i\delta(\boldsymbol{r}-\boldsymbol{r}_i)\boldsymbol{E}(\boldsymbol{r},t) + q_i\boldsymbol{v}_i\delta(\boldsymbol{r}-\boldsymbol{r}_i)\times\boldsymbol{B}(\boldsymbol{r},t)\right]dv\\ &= \int_V\left[\sum_{i=1}^{N}\rho_i(\boldsymbol{r},t)\boldsymbol{E}(\boldsymbol{r},t) + \sum_{i=1}^{N}\boldsymbol{J}_i(\boldsymbol{r},t)\times\boldsymbol{B}(\boldsymbol{r},t)\right]dv\\ &= \int_V\left[\rho(\boldsymbol{r},t)\boldsymbol{E}(\boldsymbol{r},t) + \boldsymbol{J}(\boldsymbol{r},t)\times\boldsymbol{B}(\boldsymbol{r},t)\right]dv,\end{aligned} \tag{2.11.14}$$

where

$$\boldsymbol{G}_m(t) = \sum_{i=1}^{N}\boldsymbol{G}_i(t) = \sum_{i=1}^{N}\frac{m_{i0}\boldsymbol{v}}{\sqrt{1-\left(\frac{v_i}{c}\right)^2}}. \tag{2.11.15}$$

Making use of the Maxwell equations in vacuum, (1.1.13) ~ (1.1.16), we can recast (2.11.14) into the form

$$\begin{aligned}&\frac{d}{dt}\left(\boldsymbol{G}_m + \boldsymbol{G}_f\right)\\ &= \int_V\left[\boldsymbol{E}(\nabla\cdot\varepsilon_0\boldsymbol{E}) - \varepsilon_0\boldsymbol{E}\times(\nabla\times\boldsymbol{E}) + \frac{\boldsymbol{B}}{\mu_0}(\nabla\cdot\boldsymbol{B}) - \boldsymbol{B}\times\left(\nabla\times\frac{\boldsymbol{B}}{\mu_0}\right)\right]dv,\end{aligned} \tag{2.11.16}$$

with

$$\boldsymbol{G}_f = \int_V \boldsymbol{g}_f dv = \int_V \frac{1}{c^2}\left(\boldsymbol{E}\times\frac{\boldsymbol{B}}{\mu_0}\right)dv, \tag{2.11.17}$$

where

$$\boldsymbol{g}_f = \frac{1}{c^2}\left(\boldsymbol{E} \times \frac{\boldsymbol{B}}{\mu_0}\right) = \frac{\boldsymbol{S}}{c^2}, \tag{2.11.18}$$

$\boldsymbol{G}_f$ and $\boldsymbol{g}_f$ denoting the momentum and momentum density carried by electromagnetic fields.

In order to reduce the integrand on the right-hand side of (2.11.16), we first extract the terms including field vector $\boldsymbol{E}$. With the aid of the vector identity,

$$\frac{1}{2}\nabla(\varepsilon_0 \boldsymbol{E} \cdot \boldsymbol{E}) = (\varepsilon_0 \boldsymbol{E} \cdot \nabla)\boldsymbol{E} + \varepsilon_0 \boldsymbol{E} \times (\nabla \times \boldsymbol{E}), \tag{2.11.19}$$

the terms including field vector $\boldsymbol{E}$ in (2.11.16) can be rewritten as

$$\begin{aligned} &\boldsymbol{E}(\nabla \cdot \varepsilon_0 \boldsymbol{E}) - \varepsilon_0 \boldsymbol{E} \times (\nabla \times \boldsymbol{E}) \\ &\quad = \boldsymbol{E}(\nabla \cdot \varepsilon_0 \boldsymbol{E}) + (\varepsilon_0 \boldsymbol{E} \cdot \nabla)\boldsymbol{E} - \frac{1}{2}\nabla(\varepsilon_0 E^2) \\ &\quad = \nabla \cdot \left(\varepsilon_0 \boldsymbol{E}\boldsymbol{E} - \frac{1}{2}\ddot{\boldsymbol{u}}\varepsilon_0 E^2\right), \end{aligned} \tag{2.11.20}$$

where $\ddot{\boldsymbol{u}}$ is a unit dyadic. The right-hand side of (2.11.20) is expressed in the form of the divergence of a dyadic. The terms including field vector $\boldsymbol{B}$ in (2.11.16) can also be reduced in a similar fashion.

Thus equation (2.11.16) reduces to

$$\frac{d}{dt}\left(\boldsymbol{G}_m + \boldsymbol{G}_f\right) = \int_V \nabla \cdot \ddot{\boldsymbol{T}} dv = \oint_S \ddot{\boldsymbol{T}} \cdot \boldsymbol{n} ds, \tag{2.11.21}$$

with

$$\ddot{\boldsymbol{T}} = \varepsilon_0 \boldsymbol{E}\boldsymbol{E} + \frac{1}{\mu_0}\boldsymbol{B}\boldsymbol{B} - \frac{1}{2}\ddot{\boldsymbol{u}}\left[\left(\varepsilon_0 E^2 + \frac{1}{\mu_0}B^2\right)\right], \tag{2.11.22}$$

where $\ddot{\boldsymbol{T}}$ is a dyadic expressing the Maxwell stress tensor, and $\ddot{\boldsymbol{u}}$ is expressed in terms of the unit vectors $\boldsymbol{i}_x$, $\boldsymbol{i}_y$, and $\boldsymbol{i}_z$ in the x, y, and z directions as

$$\ddot{\boldsymbol{u}} = \boldsymbol{i}_x\boldsymbol{i}_x + \boldsymbol{i}_y\boldsymbol{i}_y + \boldsymbol{i}_z\boldsymbol{i}_z. \tag{2.11.23}$$

Note that the dyadic $\ddot{\boldsymbol{T}}$ is represented in terms of components of the corresponding tensor, T_{ij} as

$$\ddot{\boldsymbol{T}} = \sum_{i=x,y,z}\sum_{j=x,y,z} \boldsymbol{i}_i\boldsymbol{i}_j T_{ij}, \tag{2.11.24}$$

from which the tensor component T_{ij} can be obtained in such a way that

$$T_{ij} = \boldsymbol{i}_i \cdot \ddot{\boldsymbol{T}} \cdot \boldsymbol{i}_j, \quad (i, j) = (x, y, z). \tag{2.11.25}$$

Specifically, the (i, j) component of the Maxwell stress tensor can be obtained from (2.11.22) as

$$T_{ij} = \varepsilon_0 E_i E_j + \frac{1}{\mu_0} B_i B_j - \frac{1}{2} \delta_{ij} \left(\varepsilon_0 E^2 + \frac{1}{\mu_0} B^2 \right), \quad (i, j) = (x, y, z), \tag{2.11.26}$$

where δ_{ij} is the Kronecker delta, being defined as

$$\delta_{ij} = \begin{cases} 1 & (i = j) \\ 0 & (i \neq j) \end{cases}. \tag{2.11.27}$$

Equation (2.11.21) represents the law of momentum conservation for a system composed of charged particles and electromagnetic fields. From (2.11.21), we recognize that the electromagnetic field carries its own momentum defined by (2.11.17) as material particles carry their mechanical momentum (2.11.15). In addition, the right-hand side of (2.11.21) denotes the force acting on a closed surface S from the outside of the surface. The integrand $\ddot{\boldsymbol{T}} \cdot \boldsymbol{n}$ in the right-hand side of (2.11.21) expresses the force normally acting per unit area on the surface S, which is the physical meaning of the Maxwell stress tensor. According to the law of momentum conservation, (2.11.21), the external force delivered through the closed surface S, which contains material particles and electromagnetic fields, supplies the temporal change of the total mechanical and field momentum in the volume V enclosed by S. If the external force acting through the surface S vanishes, the sum of the mechanical and field momentum in the volume V enclosed by S is conserved.

3. Radiation from a Moving Charged Particle

3.1 Time-Dependent Green Function for Inhomogeneous Wave Equations

We consider an inhomogeneous wave equation of the same form as treated in Sect. 1.9,

$$\nabla^2 \psi - \frac{1}{u^2}\frac{\partial^2 \psi}{\partial t^2} = -f(\boldsymbol{r},t), \tag{3.1.1}$$

where ψ corresponds to a scalar potential ϕ or any component of a vector potential $\boldsymbol{A}$, f a source producing the field ψ, and u the speed of light in an unbounded non-dispersive medium with permittivity ε and permeability μ. In order to solve the inhomogeneous wave equation (3.1.1) in an unbounded region, we will try to find a time-dependent Green function corresponding to (3.1.1), which satisfies an inhomogeneous wave equation

$$\left(\nabla^2 - \frac{1}{u^2}\frac{\partial^2}{\partial t^2}\right) G(\boldsymbol{r},t;\boldsymbol{r}',t') = -\delta(\boldsymbol{r}-\boldsymbol{r}')\delta(t-t'). \tag{3.1.2}$$

With the aid of the Green function $G(\boldsymbol{r},t;\boldsymbol{r}',t')$, the solution of (3.1.1) in an unbounded region can be represented as

$$\psi(\boldsymbol{r},t) = \int G(\boldsymbol{r},t;\boldsymbol{r}',t') f(\boldsymbol{r}',t') d\boldsymbol{r}' dt', \tag{3.1.3}$$

where $d\boldsymbol{r}' = dx'dy'dz'$.

In order to find a Green function satisfying (3.1.2), we should note that it is a function of only the difference in positions $(\boldsymbol{r}-\boldsymbol{r}')$ and the difference in times $(t-t')$. We try to solve (3.1.2) with the aid of the Fourier transforms in terms of frequency and wave numbers. First, the delta functions on the right-hand side of (3.1.2) can be expanded in the Fourier integrals in terms of frequency and wave numbers as

$$\delta(\boldsymbol{r}-\boldsymbol{r}')\delta(t-t') = \frac{1}{(2\pi)^4}\int_{-\infty}^{\infty} d\boldsymbol{k} \int_{-\infty}^{\infty} d\omega\, e^{-j\boldsymbol{k}\cdot\boldsymbol{R}} e^{j\omega\tau}, \tag{3.1.4}$$

where $\boldsymbol{R} = \boldsymbol{r} - \boldsymbol{r}'$, and $\tau = t - t'$, together with $d\boldsymbol{k} = dk_x dk_y dk_z$. Similarly, the Green function $G(\boldsymbol{r},t;\boldsymbol{r}',t')$ can also be expressed as

$$G(\boldsymbol{r},t;\boldsymbol{r}',t') = G(\boldsymbol{R},\tau) = \frac{1}{(2\pi)^4}\int_{-\infty}^{\infty} d\boldsymbol{k} \int_{-\infty}^{\infty} d\omega g(\boldsymbol{k},\omega) e^{-j\boldsymbol{k}\cdot\boldsymbol{R}} e^{j\omega\tau}. \tag{3.1.5}$$

Inserting (3.1.4) and (3.1.5) in (3.1.2), we find that the Fourier transform of the Green function, $g(\boldsymbol{k},\omega)$ is given by

$$g(\boldsymbol{k},\omega) = \frac{1}{k^2 - (\omega / u)^2}. \tag{3.1.6}$$

After the appearance of a localized point source at the time t', the disturbance represented by the Green function G is outwardly propagated as a spherical wave from the source point $\boldsymbol{r}'$ with velocity u. Hence the Green function G must satisfy the following conditions: (1) for $t < t'$, G vanishes everywhere, and (2) for $t > t'$, it represents a spherical wave traveling outwardly from the source point. The integration contours in (3.1.5) should be determined so that the above conditions imposed upon the Green function G can be met. Noting that the integrand in (3.1.5) has two poles at $\omega = \pm uk$, we can evaluate the integration in terms of ω as the Cauchy integration on the complex ω plane. Specifically, if the ω integration in (3.1.5) is to vanish everywhere for $t < t'$, we must assume that the values of ω at the poles have an infinitesimal positive imaginary part ω_i. Then, the integral (3.1.5) can be expressed as

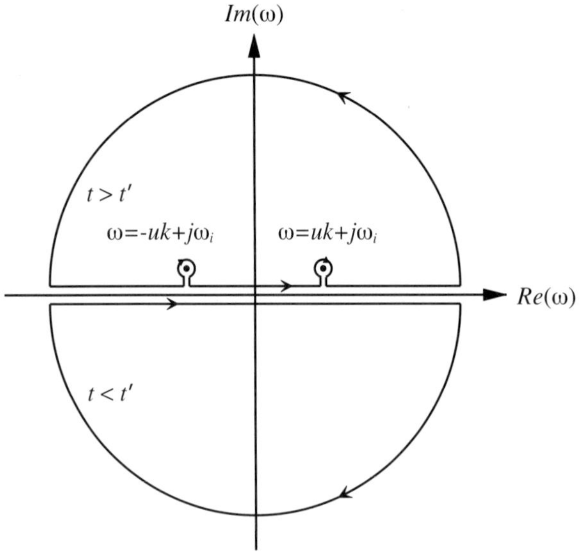

Fig. 3.1. Contours for integration on the complex ω plane

$$G(\boldsymbol{R},\tau) = \frac{1}{(2\pi)^4}\int_{-\infty}^{\infty} d\boldsymbol{k}\int_{-\infty}^{\infty} d\omega \frac{e^{j(\omega\tau - \boldsymbol{k}\cdot\boldsymbol{R})}}{k^2 - (\omega - j\omega_i)^2/u^2}. \tag{3.1.7}$$

The contour for integration in (3.1.7) is chosen as shown in Fig. 3.1. For $t < t'(\tau < 0)$, it is closed in the lower half of the complex ω plane. For $t > t'(\tau > 0)$, on the other hand, it is determined so that it is closed in the upper half of the complex ω plane, which contains two poles at $\omega = \pm uk + j\omega_i$. For $t > t'(\tau > 0)$, the ω interation in (3.1.7) can be evaluated by calculating the residues at two poles. Thus, for $\tau > 0$, we have

$$G(\boldsymbol{R},\tau) = \frac{u}{(2\pi)^3}\int_{-\infty}^{\infty} d\boldsymbol{k}\frac{\sin(u\tau k)}{k}e^{-j\boldsymbol{k}\cdot\boldsymbol{R}}. \tag{3.1.8}$$

In order to evaluate the $\boldsymbol{k}$ integration, we note the relation $d\boldsymbol{k} = 2\pi k^2 \sin\theta d\theta dk$, where θ denotes the angle between $\boldsymbol{k}$ and $\boldsymbol{R}$. Then, evaluating first the θ integration from 0 to π, we get

$$G(\boldsymbol{R},\tau) = \frac{u}{2\pi^2 R}\int_0^{\infty} dk \sin(kR)\sin(u\tau k). \tag{3.1.9}$$

Taking it into account that the integrand in (3.1.9) is an even function of k, we can rewrite (3.1.9) as

$$\begin{aligned}G(\boldsymbol{R},\tau) &= \frac{1}{8\pi^2 R}\int_{-\infty}^{\infty}\left[e^{j(R/u-\tau)uk} - e^{j(R/u+\tau)uk}\right]d(uk)\\ &= \frac{1}{8\pi^2 R}\int_{-\infty}^{\infty} e^{j(R/u-\tau)uk}d(uk)\\ &= \frac{1}{4\pi R}\delta\left(\tau - \frac{R}{u}\right).\end{aligned} \tag{3.1.10}$$

Recovering the original space and time coordinates, we get for the time-dependent Green function satisfying (3.1.2),

$$G(\boldsymbol{r},t;\boldsymbol{r}',t') = \frac{1}{4\pi|\boldsymbol{r}-\boldsymbol{r}'|}\delta\left(t - t' - \frac{|\boldsymbol{r}-\boldsymbol{r}'|}{u}\right). \tag{3.1.11}$$

With the aid of the Green function obtained above, the solution of the inhomogeneous wave equation (3.1.1) in an unbounded region can be expressed as

$$\psi(\boldsymbol{r},t) = \frac{1}{4\pi}\int \frac{f(\boldsymbol{r}',t')}{|\boldsymbol{r}-\boldsymbol{r}'|}\delta\left(t - t' - \frac{|\boldsymbol{r}-\boldsymbol{r}'|}{u}\right)d\boldsymbol{r}'dt'. \tag{3.1.12}$$

Evaluating the integration in terms of t', we find

$$\psi(\boldsymbol{r},t) = \frac{1}{4\pi}\int \frac{1}{|\boldsymbol{r}-\boldsymbol{r}'|} f\left(\boldsymbol{r}', t - \frac{|\boldsymbol{r}-\boldsymbol{r}'|}{u}\right) d\boldsymbol{r}', \tag{3.1.13}$$

which is the same as obtained in (1.9.13).

Applying the general solution (3.1.12) to the inhomogeneous wave equations for vector and scalar potentials, (1.5.10) and (1.5.11), we get for their solutions,

$$\boldsymbol{A}(\boldsymbol{r},t) = \frac{\mu}{4\pi}\int \frac{\boldsymbol{J}(\boldsymbol{r}',t')}{|\boldsymbol{r}-\boldsymbol{r}'|} \delta\left(t - t' - \frac{|\boldsymbol{r}-\boldsymbol{r}'|}{u}\right) d\boldsymbol{r}'dt', \tag{3.1.14}$$

$$\phi(\boldsymbol{r},t) = \frac{1}{4\pi\varepsilon}\int \frac{\rho(\boldsymbol{r}',t')}{|\boldsymbol{r}-\boldsymbol{r}'|} \delta\left(t - t' - \frac{|\boldsymbol{r}-\boldsymbol{r}'|}{u}\right) d\boldsymbol{r}'dt'. \tag{3.1.15}$$

From these equations, we can obtain the solutions for vector and scalar potentials produced by current and charge densities which are arbitrary functions of position and time.

3.2 Liénard-Wiechert Potentials

Let us try to find vector and scalar potentials produced by a charged particle moving with an arbitrary velocity in an unbounded medium with permittivity ε and permeability μ. For this purpose, we assume that a particle with electric charge q is moving with the velocity $\boldsymbol{v}(t)$ and occupying the position $\boldsymbol{r}_0(t)$ at a particular instant of time, t. For the time being, the magnitude of the velocity of a charged particle is assumed to be smaller than the speed of light in an unbounded medium with permittivity ε and permeability μ. For the vacuum case, this condition is always satisfied. In a later section, we will discuss in detail the case where the velocity of a charged particle exceeds the velocity of light in a material medium.

The current and charge densities for a charged particle specified above can be expressed as

$$\boldsymbol{J}(\boldsymbol{r},t) = q\boldsymbol{v}(t)\delta[\boldsymbol{r}-\boldsymbol{r}_0(t)], \tag{3.2.1}$$

$$\rho(\boldsymbol{r},t) = q\delta[\boldsymbol{r}-\boldsymbol{r}_0(t)]. \tag{3.2.2}$$

Inserting these expressions for current and charge densities due to a moving charged particle in (3.1.14) and (3.1.15), we have

$$\boldsymbol{A}(\boldsymbol{r},t) = \frac{\mu q}{4\pi}\int \frac{\boldsymbol{v}(t')\delta[\boldsymbol{r}'-\boldsymbol{r}_0(t')]}{|\boldsymbol{r}-\boldsymbol{r}'|} \delta\left(t - t' - \frac{|\boldsymbol{r}-\boldsymbol{r}'|}{u}\right) d\boldsymbol{r}'dt', \tag{3.2.3}$$

$$\phi(\boldsymbol{r},t) = \frac{q}{4\pi\varepsilon}\int \frac{\delta[\boldsymbol{r}' - \boldsymbol{r}_0(t')]}{|\boldsymbol{r} - \boldsymbol{r}'|}\delta\left(t - t' - \frac{|\boldsymbol{r} - \boldsymbol{r}'|}{u}\right)d\boldsymbol{r}'dt'. \tag{3.2.4}$$

The spatial integration in (3.2.3) and (3.2.4) can readily be carried out with the results,

$$\boldsymbol{A}(\boldsymbol{r},t) = \frac{\mu q}{4\pi}\int \frac{\boldsymbol{v}(t')}{R(t')}\delta\left(t - t' - \frac{R(t')}{u}\right)dt', \tag{3.2.5}$$

$$\phi(\boldsymbol{r},t) = \frac{q}{4\pi\varepsilon}\int \frac{1}{R(t')}\delta\left(t - t' - \frac{R(t')}{u}\right)dt', \tag{3.2.6}$$

where

$$R(t') = |\boldsymbol{r} - \boldsymbol{r}_0(t')|, \tag{3.2.7}$$

which denotes the distance between a particular field point and the retarded position of a charged particle at a particular retarded time.

In order to further reduce the integrals on the right hand sides of (3.2.5) and (3.2.6), we note the following relation concerning the integral of the delta function:

$$\int f(\xi)\delta[\alpha(\xi) - \alpha_0]d\xi = \left.\frac{f(\xi)}{d\alpha / d\xi}\right|_{\alpha(\xi)=\alpha_0}, \tag{3.2.8}$$

where $f(\xi)$ is an arbitrary continuous function of ξ, and α_0 an arbitrary constant. To evaluate the integrals (3.2.5) and (3.2.6) with the aid of the relation (3.2.8), we first note the following relation:

$$\frac{d}{dt'}\left[t' + \frac{R(t')}{u}\right] = 1 - \frac{1}{u}\boldsymbol{i}_{t'} \cdot \boldsymbol{v} \equiv \eta, \tag{3.2.9}$$

where $\boldsymbol{i}_{t'}$ denotes a unit vector directed from the retarded position of a charged particle to a particular field point. In deriving the relation (3.2.9), we used the relations $R(t') = r - \boldsymbol{i}_{t'} \cdot \boldsymbol{r}_0(t')$, which is obtained from (3.2.7), and $d\boldsymbol{r}_0 / dt' = \boldsymbol{v}$.

Using the relations (3.2.8) and (3.2.9) in the integrals (3.2.5) and (3.2.6), we get for the final results for vector and scalar potentials produced by a moving charged particle,

$$\boldsymbol{A}(\boldsymbol{r},t) = \frac{\mu q}{4\pi}\frac{\boldsymbol{v}}{\eta R}, \tag{3.2.10}$$

$$\phi(\boldsymbol{r},t) = \frac{q}{4\pi\varepsilon}\frac{1}{\eta R}, \tag{3.2.11}$$

in which the values of R, $\boldsymbol{v}$, and η are evaluated at the retarded time determined from the following relation:

$$t' = t - \frac{R(t')}{u}. \tag{3.2.12}$$

The electromagnetic potentials given by (3.2.10) and (3.2.11) are referred to as the Liénard-Wiechert potentials.

3.3 Fields Produced by a Moving Charged Particle

With the aid of the Liénard-Wiechert potentials, we calculate the electromagnetic fields produced by a moving charged particle. Differentiating the electromagnetic potentials given by (3.2.10) and (3.2.11) with respect to space and time, we can find the electromagnetic fields due to a moving charged particle as

$$\boldsymbol{E} = -\nabla\phi - \frac{\partial \boldsymbol{A}}{\partial t}, \tag{3.3.1}$$

$$\boldsymbol{B} = \nabla \times \boldsymbol{A}. \tag{3.3.2}$$

In treating the fields produced by a moving charged particle, we should carefully examine the physical meaning of spatial and temporal differentiations in (3.3.1) and (3.3.2). At the outset of our discussion, we have assumed that the position and velocity of a charged particle are given as functions of the retarded time t' but not of the time of observation, t. In addition, the differential operator ∇ means spatial partial differentiation with respect to the coordinates of the field point, with the time of observation, not the retarded time, kept invariant. Then, we should keep it in mind that partial differetiation with respect to spatial coordinates compares potentials at positions separated by infinitesimal distance and these potentials were emitted from a charged particle at different retarded times. Hence the temporal and spatial derivatives in (3.3.1) and (3.3.2) must be rewritten with the aid of the temporal derivative with respect to the retarded time, taking the conditions described above into account [3.2].

To find the appropriate transformation for the temporal and spatial derivatives, we note from (3.2.7) that the distance between a particular position of observation and the retarded position of a charged particle, R is a function of the position of observation $\boldsymbol{r}$ and the retarded time t'. Hence R is expressed as

$$R = R(\boldsymbol{r}, t') = u(t - t'). \tag{3.3.3}$$

First, differentiating R with respect to t, we get

$$\frac{\partial R}{\partial t} = \frac{\partial R}{\partial t'}\frac{\partial t'}{\partial t} = u\left(1 - \frac{\partial t'}{\partial t}\right). \tag{3.3.4}$$

Using the relation $\partial R / \partial t' = -\boldsymbol{i}_{t'} \cdot \boldsymbol{v}$, which is obtained from (3.2.7), we find, from (3.3.4), the following relation:

$$\frac{\partial t'}{\partial t} = \frac{1}{1 - \boldsymbol{i}_{t'} \cdot \boldsymbol{v}/u} = \frac{1}{\eta}, \tag{3.3.5}$$

from which we get for the proper transformation for the time derivatives,

$$\frac{\partial}{\partial t} = \frac{1}{\eta}\frac{\partial}{\partial t'}. \tag{3.3.6}$$

Next, let us try to find the transformation for the spatial derivatives. For this purpose, we impose the vector operator ∇ on (3.3.3), with the time of observation t fixed. Then, we have

$$\nabla R = \nabla_{t'} R + \frac{\partial R}{\partial t'} \nabla t' = -u \nabla t', \tag{3.3.7}$$

where $\nabla_{t'}$ denotes the spatial vector operator with the retarded time fixed. Noting the relations $\nabla_{t'} R = \boldsymbol{i}_{t'}$ and $\partial R / \partial t' = -\boldsymbol{i}_{t'} \cdot \boldsymbol{v}$, we get

$$\nabla t' = -\frac{\boldsymbol{i}_{t'}}{u\eta}. \tag{3.3.8}$$

Using (3.3.8) in (3.3.7), we have for the transformation of the vector spatial operators,

$$\nabla = \nabla_{t'} - \frac{\boldsymbol{i}_{t'}}{u\eta}\frac{\partial}{\partial t'}. \tag{3.3.9}$$

Applying the temporal and spatial differential operators (3.3.6) and (3.3.9) to (3.3.1) and (3.3.2), we get for the electromagnetic fields produced by a moving charged particle,

$$\boldsymbol{E} = \frac{q}{4\pi\varepsilon}\frac{1}{\gamma_n^2 \eta^3 R^2}\left(\boldsymbol{i}_{t'} - \frac{\boldsymbol{v}}{u}\right) + \frac{\mu u q}{4\pi}\frac{1}{\eta^3 R}\left\{\boldsymbol{i}_{t'} \times \left[\left(\boldsymbol{i}_{t'} - \frac{\boldsymbol{v}}{u}\right) \times \frac{\dot{\boldsymbol{v}}}{u}\right]\right\}, \tag{3.3.10}$$

$$\begin{aligned}\boldsymbol{B} &= \frac{\mu q}{4\pi}\frac{1}{\gamma_n^2 \eta^3 R^2}(\boldsymbol{v} \times \boldsymbol{i}_{t'}) + \frac{\mu q}{4\pi}\frac{1}{\eta^3 R}\left[(\boldsymbol{i}_{t'} \cdot \frac{\dot{\boldsymbol{v}}}{u})(\frac{\boldsymbol{v}}{u} \times \boldsymbol{i}_{t'}) + \eta(\frac{\dot{\boldsymbol{v}}}{u} \times \boldsymbol{i}_{t'})\right] \\ &= \frac{1}{u}(\boldsymbol{i}_{t'} \times \boldsymbol{E}),\end{aligned} \tag{3.3.11}$$

where

$$\gamma_n = \frac{1}{\sqrt{1 - n^2\beta^2}}, \quad n = \frac{c}{u}, \quad \beta = \frac{v}{c}, \quad \dot{\boldsymbol{v}} = \frac{d\boldsymbol{v}}{dt'}. \tag{3.3.12}$$

c and n denoting the speed of light in vacuum and the refractive index of the medium.

As seen from (3.3.10) and (3.3.11), the electromagnetic fields produced by a moving charged particle are composed of two different parts. The first term in (3.3.10) and the corresponding term in (3.3.11) decay as $1/R^2$ for large distances. These terms are formally identical to the fields due to a uniformly moving charge, which will be discussed in the next section, and do not contribute to radiation. On the other hand, the second term in (3.3.10) and the corresponding term in (3.3.11) decay as $1/R$ for large distances, contributing to radiation. In order to discuss the properties of nonradiating terms and radiation fields in more detail, let us investigate the physical meaning of the vector appearing in (3.3.10) and (3.3.11),

$$\boldsymbol{R}_t = R\left(\boldsymbol{i}_{t'} - \frac{\boldsymbol{v}}{u}\right) = \boldsymbol{R} - \frac{R}{u}\boldsymbol{v}. \tag{3.3.13}$$

The second term on the right-hand side of (3.3.13) expresses the distance a charged particle would travel from its retarded position by the time of observation if it had uniformly moved with the velocity at the retarded time. Hence $\boldsymbol{R}_t$ denotes the radius vector of an observation point as referred to the virtual present position of a charged particle. Thus the nonradiating electric field, namely, the first term in (3.3.10) is directed from the virtual present position to the point of observation. For radiation fields, on the other hand, the electric and magnetic fields are perpendicular to each other, and also to the direction from the retarded position of a charged particle to the point of observation. In addition, the ratio of the magnitudes of electric and magnetic fields is equal to the wave impedance of the medium. Therefore, we find that the radiation fields have the same properties as plane waves.

As is evident from the second term in (3.3.10), a charged particle must generally undergo an accelerated motion so that it can radiate electromagnetic energy. For the special case, a charged particle uniformly moving in an material medium can also contribute to radiation if its velocity is greater than the light velocity in that medium. This radiation is called the Cherenkov radiation, which will be discussed in detail in a later section.

3.4 Fields of a Charged Particle in Uniform Motion

The electric and magnetic fields produced by a charged particle in uniform motion are given by the first terms in (3.3.10) and (3.3.11),

$$\boldsymbol{E} = \frac{q}{4\pi\varepsilon}\frac{1}{\gamma_n^2\eta^3R^2}\left(\boldsymbol{i}_{t'} - \frac{\boldsymbol{v}}{u}\right), \tag{3.4.1}$$

$$\boldsymbol{B} = \frac{\mu q}{4\pi}\frac{1}{\gamma_n^2\eta^3R^2}(\boldsymbol{v}\times\boldsymbol{i}_{t'}). \tag{3.4.2}$$

These fields are expressed in reference to the retarded position of a charged particle. For a uniformly moving charge, these fields can also be represented relative to the present position of the charged particle. By the present position, we mean the point which a charged particle occupies when the signal emitted at the retarded position of the particle arrives at the field point.

In order to express the electric and magnetic fields (3.4.1) and (3.4.2) relative to the present position of the charged particle, we first rewrite ηR appearing in the denominators of the field expressions. From the definition of η in (3.2.9), we have

$$(\eta R)^2 = R^2\left(1 - \frac{1}{u}\boldsymbol{i}_{t'} \cdot \boldsymbol{v}\right)^2 = R^2(1 - n\beta\cos\theta)^2, \tag{3.4.3}$$

where θ is the angle between $\boldsymbol{i}_{t'}$ and $\boldsymbol{v}$. To modify the right-hand side of (3.4.3), we note the relation

$$R_t^2 = R^2\left[1 - 2n\beta\cos\theta + (n\beta)^2\right], \tag{3.4.4}$$

which is obtained from (3.3.13). In addition, referring to Fig. 3.2, we find

$$R\sin\theta = R_t\sin\theta_t, \tag{3.4.5}$$

where θ_t denotes the angle between $\boldsymbol{R}_t$ and $\boldsymbol{v}$. Using the relations (3.4.4) and (3.4.5) in (3.4.3), we get

$$\eta R = R_t\sqrt{1 - n^2\beta^2\sin^2\theta_t}\ . \tag{3.4.6}$$

Inserting the relations (3.3.13) and (3.4.6) in (3.4.1) and (3.4.2), we finally get for the electric and magnetic fields due to a charged particle in uniform motion, which are represented relative to the present position of the charge,

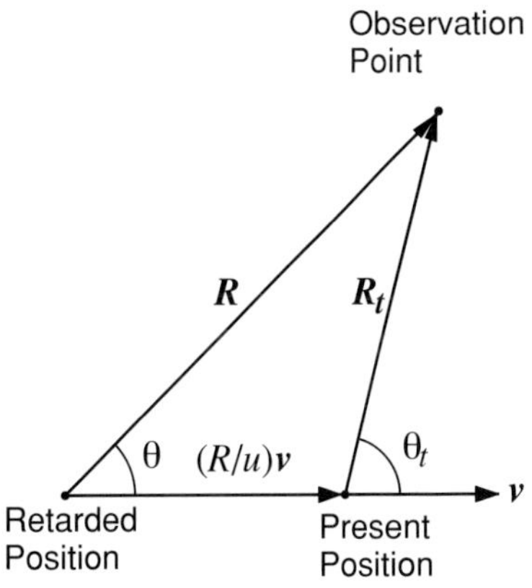

Fig. 3.2. Retarded and present positions of a uniformly moving charged particle

$$\boldsymbol{E} = \boldsymbol{i}_t \frac{q}{4\pi\varepsilon R_t^2} \frac{1}{\gamma_n^2\left[1 - n^2\beta^2 \sin^2\theta_t\right]^{3/2}}, \tag{3.4.7}$$

$$\boldsymbol{B} = \frac{\mu q}{4\pi R_t^2} \frac{1}{\gamma_n^2\left[1 - n^2\beta^2 \sin^2\theta_t\right]^{3/2}} (\boldsymbol{v} \times \boldsymbol{i}_t) = \frac{1}{u^2}(\boldsymbol{v} \times \boldsymbol{E}), \tag{3.4.8}$$

where R_t is the distance between the present position of the charged particle and the field point, and we assume that the condition $n\beta < 1$ holds or the velocity of the charged particle is smaller than that of light in a material medium. The case of $n\beta >$ 1 will be discussed in a later section. For small velocities $(\beta \to 0)$, Eq. (3.4.7) reduces to the Coulomb field, and Eq. (3.4.8) to the field obtained from the Ampere law. As $n\beta$ tends to unity, the fields given by (3.4.7) and (3.4.8) approach transverse fields, which get contained in the plane perpendicular to the velocity of the charged particle. However, it should be noted that these fields do not radiate electromagnetic energy because they decay as $1/R^2$ for large distances.

3.5 Radiation Fields of a Charged Particle in Accelerated Motion

The electromagnetic fields radiated from a charged particle in accelerated motion are extracted from (3.3.10) and (3.3.11) as

$$\boldsymbol{E}_{rad} = \frac{\mu u q}{4\pi} \frac{1}{\eta^3 R} \left\{ \boldsymbol{i}_{t'} \times \left[\left(\boldsymbol{i}_{t'} - \frac{\boldsymbol{v}}{u} \right) \times \frac{\dot{\boldsymbol{v}}}{u} \right] \right\}, \tag{3.5.1}$$

$$\boldsymbol{B}_{rad} = \frac{1}{u}(\boldsymbol{i}_{t'} \times \boldsymbol{E}_{rad}), \tag{3.5.2}$$

from which we get the Poynting vector for the radiation fields of an accelerated charge,

$$\boldsymbol{S} = \boldsymbol{E}_{rad} \times \frac{\boldsymbol{B}_{rad}}{\mu} = \boldsymbol{i}_{t'} \left(\frac{q}{4\pi} \right)^2 \sqrt{\frac{\mu}{\varepsilon}} \frac{1}{\eta^6 R^2} \left| \boldsymbol{i}_{t'} \times \left[\left(\boldsymbol{i}_{t'} - \frac{\boldsymbol{v}}{u} \right) \times \frac{\dot{\boldsymbol{v}}}{u} \right] \right|^2 . \tag{3.5.3}$$

As seen from (3.5.3), the Poynting vector has only the radial component as referred to the retarded position of a charged particle. Let the amount of energy radiated from a charged particle in the retarded time interval dt' in a solid angle $d\Omega$ be denoted by $dP\,dt'$. Then, this same amount of energy traverses a small area $R^2 d\Omega$ around the observation point on a large sphere of radius R in the observation time interval dt. Hence we find the following relation:

$$dPdt' = (\boldsymbol{i}_{t'} \cdot \boldsymbol{S})R^2 d\Omega dt, \tag{3.5.4}$$

from which we have

$$\frac{dP}{d\Omega} = (\boldsymbol{i}_{t'} \cdot \boldsymbol{S}) R^2 \frac{dt}{dt'}. \tag{3.5.5}$$

Then, with the aid of (3.5.3) and (3.3.5), we get for the amount of energy radiated from an accelerated charge in a unit solid angle per unit time at the retarded position of the charge,

$$\frac{dP}{d\Omega} = \left(\frac{q}{4\pi}\right)^2 \sqrt{\frac{\mu}{\varepsilon}} \frac{1}{\eta^5} \left| \boldsymbol{i}_{t'} \times \left[\left(\boldsymbol{i}_{t'} - \frac{\boldsymbol{v}}{u} \right) \times \frac{\dot{\boldsymbol{v}}}{u} \right] \right|^2. \tag{3.5.6}$$

Equation (3.5.6) expresses the angular distribution of the power radiated from an accelerated charge. For the special case where the acceleration and velocity of a charged particle are parallel to each other, an explicit form of (3.5.6) is given by

$$\frac{dP}{d\Omega} = \left(\frac{q}{4\pi}\right)^2 \sqrt{\frac{\mu}{\varepsilon}} \left(\frac{\dot{v}}{u}\right)^2 \frac{\sin^2\theta}{(1 - n\beta\cos\theta)^5}, \tag{3.5.7}$$

where θ denotes the angle between the directions of observation and the velocity of the charged particle (the angle between $\boldsymbol{i}_{t'}$ and $\boldsymbol{v}$). For another important case where the acceleration and velocity of a charged particle are perpendicular to each other, Eq. (3.5.6) can be rewritten explicitly in the form

$$\frac{dP}{d\Omega} = \left(\frac{q}{4\pi}\right)^2 \sqrt{\frac{\mu}{\varepsilon}} \frac{(\dot{v}/u)^2}{(1 - n\beta\cos\theta)^3} \left[1 - \frac{\sin^2\theta\cos^2\varphi}{\gamma_n^2 (1 - n\beta\cos\theta)^2} \right], \tag{3.5.8}$$

where φ is the angle between the acceleration of the charge and the projection of $\boldsymbol{i}_{t'}$ on the plane perpendicular to the velocity of the charge. In Fig. 3.3, we show the angular distribution of the power radiated for the two cases of the acceleration of the charged particle parallel and perpendicular to its velocity. From this figure, we see that the radiation is concentrated around the direction of the velocity of the charged particle for large values of $n\beta$. In general, the radiation is confined within a small angle ($\Delta\theta \approx 1/\gamma_n$) around the direction of the velocity of the charged particle for large values of $n\beta$.

The total energy radiated from a charged particle per unit time can be obtained by integrating (3.5.6) over the whole solid angle. After some manipulations, we find for the total power radiated from the charge at the retarded time,

$$P(t') = \gamma_n^6 \frac{q^2}{6\pi} \sqrt{\frac{\mu}{\varepsilon}} \left[\left(\frac{\dot{\boldsymbol{v}}}{u}\right)^2 - \left(\frac{\boldsymbol{v}}{u} \times \frac{\dot{\boldsymbol{v}}}{u}\right)^2 \right]. \tag{3.5.9}$$

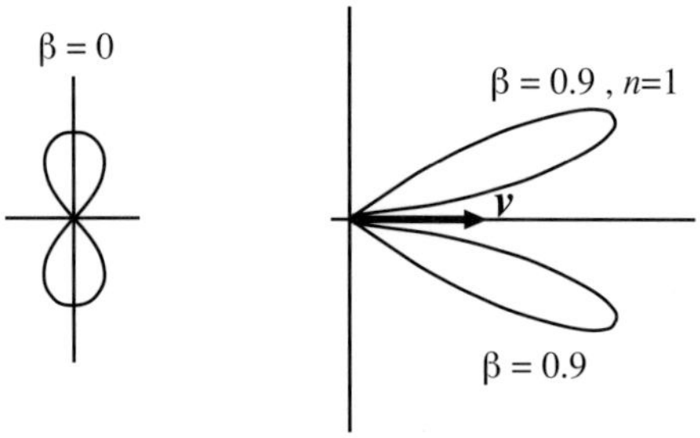

(a) $\boldsymbol{v} \,//\, \dot{\boldsymbol{v}}$

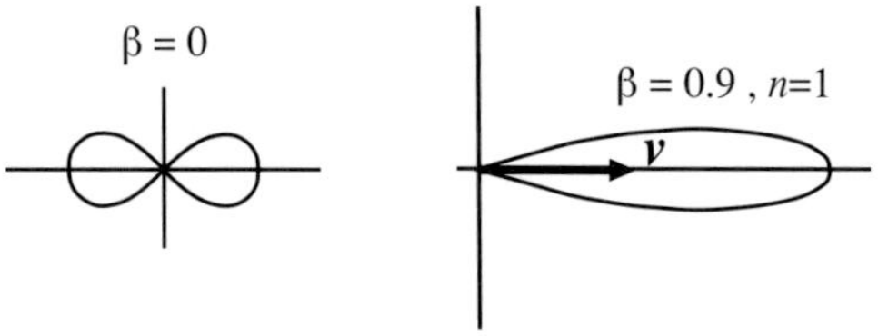

(b) $\boldsymbol{v} \perp \dot{\boldsymbol{v}}$

Fig. 3.3. Angular distributions of power radiated from an accelerated charge. The velocity and acceleration of the charge are parallel in (a), and perpendicular in (b).

For the special case of $\dot{\boldsymbol{v}}$ parallel to $\boldsymbol{v}$, $P(t')$ takes the form

$$P_{//}(t') = \gamma_n^6 \frac{q^2}{6\pi} \sqrt{\frac{\mu}{\varepsilon}} \left(\frac{\dot{v}}{u}\right)^2 . \tag{3.5.10}$$

For the case of $\dot{\boldsymbol{v}}$ perpendicular to $\boldsymbol{v}$, on the other hand, we find

$$P_{\perp}(t') = \gamma_n^4 \frac{q^2}{6\pi} \sqrt{\frac{\mu}{\varepsilon}} \left(\frac{\dot{v}}{u}\right)^2 . \tag{3.5.11}$$

Finally, let us consider the special case of $n\beta << 1$. For this case, we have

$$\frac{dP}{d\Omega} = \left(\frac{q}{4\pi}\right)^2 \sqrt{\frac{\mu}{\varepsilon}} \left(\frac{\dot{v}}{u}\right)^2 \sin^2 \xi, \tag{3.5.12}$$

$$P(t') = \frac{q^2}{6\pi} \sqrt{\frac{\mu}{\varepsilon}} \left(\frac{\dot{v}}{u}\right)^2 , \tag{3.5.13}$$

where ξ denotes the angle between the directions of observation and the acceleration of the charge. For the radiation from a charged particle in simple harmonic motion, Eqs. (3.5.12) and (3.5.13) coincide with the results obtained for the radiation from an oscillating electric dipole, (1.10.44) and (1.10.43).

3.6 Frequency Spectrum of the Radiated Energy

Let us assume that a charged particle moving with an extremely relativistic velocity is accelerated for a finite duration of time and the total energy radiated from the charge is bounded. Then, the total energy radiated from a moving charge per unit solid angle as viewed from the field point is represented with the aid of (3.5.3) as

$$\frac{dU}{d\Omega} = \int_{-\infty}^{\infty} (\boldsymbol{i}_{t'} \cdot \boldsymbol{S}) R^2 dt = \int_{-\infty}^{\infty} |f(t)|^2 dt, \tag{3.6.1}$$

where t denotes the time at the field point and $f(t)$ is given by

$$f(t) = \frac{q}{4\pi}\sqrt{\mu u}\,\frac{1}{\eta^3}\left\{\boldsymbol{i}_{t'} \times \left[\left(\boldsymbol{i}_{t'} - \frac{\boldsymbol{v}}{u}\right) \times \frac{\dot{\boldsymbol{v}}}{u}\right]\right\}. \tag{3.6.2}$$

In view of Parseval's theorem given by (1.10.33), we can express Eq. (3.6.1) in terms of the Fourier transform of $f(t)$ as well,

$$\frac{dU}{d\Omega} = \frac{1}{\pi}\int_0^{\infty} \left|\tilde{f}(\omega)\right|^2 d\omega, \tag{3.6.3}$$

where $\tilde{f}(\omega)$ denotes the Fourier transform of $f(t)$.

In our definition, the Fourier transform of $f(t)$ takes the form

$$\begin{aligned}\tilde{f}(\omega) &= \int_{-\infty}^{\infty} f(t) e^{-j\omega t} dt \\ &= \frac{q}{4\pi}\sqrt{\mu u}\int_{-\infty}^{\infty} \frac{1}{\eta^3}\left\{\boldsymbol{i}_{t'} \times \left[\left(\boldsymbol{i}_{t'} - \frac{\boldsymbol{v}}{u}\right) \times \frac{\dot{\boldsymbol{v}}}{u}\right]\right\} e^{-j\omega t} dt.\end{aligned} \tag{3.6.4}$$

Transforming the integration in terms of t to that in terms of t', with the aid of the relations (3.2.12) and (3.3.5), we get

$$\tilde{f}(\omega) = \frac{q}{4\pi}\sqrt{\mu u}\int_{-\infty}^{\infty} \frac{1}{\eta^2}\left\{\boldsymbol{i}_{t'} \times \left[\left(\boldsymbol{i}_{t'} - \frac{\boldsymbol{v}}{u}\right) \times \frac{\dot{\boldsymbol{v}}}{u}\right]\right\} e^{-j\omega[t' + R(t')/u]} dt', \tag{3.6.5}$$

where t' denotes the time at the retarded position of the charged particle. We assume that the charged particle moves only by a small distance compared with the distance to the observation point while it is undergoing an accelerated motion. Then, we can regard the unit vector $\boldsymbol{i}_{t'}$, which is directed from the retarded position of the charge to the observation point, as approximately constant in the integrand of

(3.6.5). We also have the relation $R(t') = r - \boldsymbol{i}_{t'} \cdot \boldsymbol{r}_0(t')$, which is obtained from (3.2.7). Taking these approximations into account, we can rewrite (3.6.5) as

$$\tilde{f}(\omega) = \frac{q}{4\pi}\sqrt{\mu u}\, e^{-j(\omega/u)r} \int_{-\infty}^{\infty} \frac{1}{\eta^2}\left\{\boldsymbol{i}_{t'} \times \left[\left(\boldsymbol{i}_{t'} - \frac{\boldsymbol{v}}{u}\right) \times \frac{\dot{\boldsymbol{v}}}{u}\right]\right\} e^{-j\omega[t' - \boldsymbol{i}_{t'} \cdot \boldsymbol{r}_0(t')/u]} dt'. \tag{3.6.6}$$

Noting that the following relation holds in the integrand of (3.6.6):

$$\frac{1}{\eta^2}\left\{\boldsymbol{i}_{t'} \times \left[\left(\boldsymbol{i}_{t'} - \frac{\boldsymbol{v}}{u}\right) \times \frac{\dot{\boldsymbol{v}}}{u}\right]\right\} = \frac{d}{dt'}\left[\frac{\boldsymbol{i}_{t'} \times \left(\boldsymbol{i}_{t'} \times \frac{\boldsymbol{v}}{u}\right)}{\eta}\right], \tag{3.6.7}$$

we can carry out the integration in (3.6.6) by parts to get

$$\tilde{f}(\omega) = \frac{j\omega q}{4\pi}\sqrt{\mu u}\, e^{-j(\omega/u)r} \int_{-\infty}^{\infty} \left[\boldsymbol{i}_{t'} \times \left(\boldsymbol{i}_{t'} \times \frac{\boldsymbol{v}}{u}\right)\right] e^{-j\omega[t' - \boldsymbol{i}_{t'} \cdot \boldsymbol{r}_0(t')/u]} dt'. \tag{3.6.8}$$

Inserting (3.6.8) in (3.6.3), we have for the total energy radiated from an accelerated charge per unit solid angle,

$$\frac{dU}{d\Omega} = \frac{q^2}{16\pi^3}\sqrt{\frac{\mu}{\varepsilon}} \int_0^{\infty} \omega^2 \left|\int_{-\infty}^{\infty} \left[\boldsymbol{i}_{t'} \times \left(\boldsymbol{i}_{t'} \times \frac{\boldsymbol{v}}{u}\right)\right] e^{-j\omega[t' - \boldsymbol{i}_{t'} \cdot \boldsymbol{r}_0(t')/u]} dt'\right|^2 d\omega, \tag{3.6.9}$$

which is expressed in the form

$$\frac{dU}{d\Omega} = \int_0^{\infty} \frac{d^2 F(\omega, \boldsymbol{i}_{t'})}{d\Omega d\omega} d\omega, \tag{3.6.10}$$

$$\frac{d^2 F(\omega, \boldsymbol{i}_{t'})}{d\Omega d\omega} = \frac{q^2}{16\pi^3}\sqrt{\frac{\mu}{\varepsilon}}\, \omega^2 \left|\int_{-\infty}^{\infty} \left[\boldsymbol{i}_{t'} \times \left(\boldsymbol{i}_{t'} \times \frac{\boldsymbol{v}}{u}\right)\right] e^{-j\omega[t' - \boldsymbol{i}_{t'} \cdot \boldsymbol{r}_0(t')/u]} dt'\right|^2. \tag{3.6.11}$$

Equation (3.6.11) represents the energy radiated from an accelerated charge per unit solid angle per unit frequency interval, namely, the frequency spectrum of the energy radiated per unit solid angle.

3.7 Synchrotron Radiation

Synchrotron radiation is one of the most interesting topics from a viewpoint of applications of radiation from charged particles moving with relativistic velocities [3.4]. Before entering the discussion of synchrotron radiation, let us first consider why it is so important in the radiation from a relativistically moving charged particle. For this purpose, we compare two typical cases of motion where the acceleration and velocity of a charged particle are parallel and perpendicular to each other under the influence of external forces with the same magnitude. We should note that an arbitrary motion of a charged particle can be decomposed into the above two cases.

First, for one case of motion where the acceleration and velocity of a charged particle are parallel to each other, the acceleration of the charge can be obtained from (2.9.1), (2.9.2) and (2.9.10),

$$\dot{v} = \frac{1}{\gamma^3}\frac{F}{m_0}, \tag{3.7.1}$$

with

$$\gamma = \frac{1}{\sqrt{1-\beta^2}}, \qquad \beta = \frac{v}{c}, \tag{3.7.2}$$

where F and m_0 denote the magnitude of an external force applied to the charge and the rest mass of the charge, respectively. Inserting (3.7.1) in (3.5.10), we get for the total power radiated,

$$P_{//}(t') = \frac{q^2F^2}{6\pi\varepsilon_0 c^3 m_0^2}, \tag{3.7.3}$$

where we assumed $n = 1$ or $u = c$ in (3.5.10) to consider the radiation in vacuum.

On the other hand, for the other case where the acceleration and velocity of a charged particle are perpendicular to each other, we obtain for the acceleration of the charge,

$$\dot{v} = \frac{1}{\gamma}\frac{F}{m_0}. \tag{3.7.4}$$

Substituting (3.7.4) into (3.5.11), we have for the total power radiated,

$$P_{\perp}(t') = \gamma^2\frac{q^2F^2}{6\pi\varepsilon_0 c^3 m_0^2}. \tag{3.7.5}$$

Comparing the results (3.7.3) and (3.7.5), we find that the total power radiated from a relativistically moving charged particle under the influence of an external force with the same magnitude is greater by the factor γ^2 for the latter case than for the former case. For example, for $\gamma = 100$, the radiation for the latter case is 10^4 times more intense than for the former case. The former case corresponds to the radiation emitted when a charge is accelerated or decelerated along the direction of its motion. On the other hand, the latter case represents the radiation emitted when the direction of motion of a charged particle is abruptly changed by some transverse force. As seen from a specific example illustrated above, the latter case is more important and interesting for the radiation from a charged particle moving with extremely relativistic velocity. The radiation emitted when the direction of motion of a charged particle moving with extremely relativistic velocity is abruptly changed is generally referred to as synchrotron radiation. The orbit of a charged particle can be easily curved by the Lorentz force due to a static magnetic field. In the following discussion, we treat only a charged particle moving in vacuum.

The radiation from a charged particle moving in an arbitrary direction with extremely relativistic velocity can be regarded as a coherent superposition of radiation obtained in two cases where the acceleration and velocity of the charge are parallel and perpendicular to each other. As is evident from the foregoing discussion, for an extremely relativistic case, the contribution to radiation from the acceleration component of the charge parallel to its velocity can be neglected as compared with that from the acceleration component perpendicular to it. Hence the radiation from a charged particle moving arbitrarily with extremely relativistic velocity around a particular instant of time is approximately equal to the radiation emitted by a charge moving along a small circular arc (see Fig. 3.4). Remember that the acceleration and velocity for a circular motion are perpendicular to each other. The radius of curvature for a small circular arc, ρ is given by

$$\rho = \frac{v^2}{\dot{v}_\perp}, \tag{3.7.6}$$

where v and $\dot{v}_\perp$ denote the instantaneous values of the charge velocity and the acceleration component perpendicular to it. The relation (3.7.6) can be easily obtained by noting the relations $v = \Omega\rho$ and $\dot{v}_\perp = \Omega^2\rho$, where Ω denotes the instantaneous value of the angular frequency of rotation. The angular distribution of the power radiated from a charge moving along a small circular arc is given by

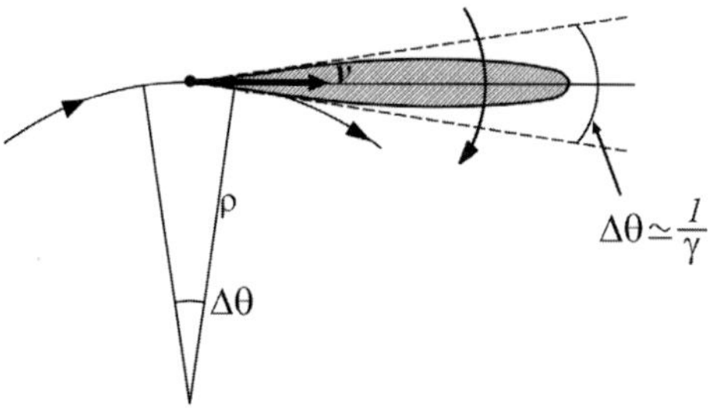

Fig. 3.4. Radiation from a charged particle moving along a circular arc with radius of curvature ρ

(3.5.8). The power radiated is confined within a small angle ($\Delta\theta \approx 1/\gamma$) around the direction of the instantaneous velocity of the charge at a particular instant of time. This last statement can be proved in the following manner. First, we note that the angular distribution of the power radiated, (3.5.8) becomes maximum for $\varphi = \pi/2$ and $\theta = 0$. Then, expanding $\cos\theta$ in (3.5.8) in a Taylor series around $\theta = 0$ with φ kept equal to $\pi/2$, we get for sufficiently large values of β,

$$\frac{dP}{d\Omega} = \frac{1}{2}\left(\frac{q}{\pi}\right)^2 \sqrt{\frac{\mu_0}{\varepsilon_0}} \frac{\gamma^6 (\dot{v}/c)^2}{(1+\gamma^2\theta^2)^3}, \tag{3.7.7}$$

where $n = 1$ or $u = c$ are assumed. For the angular width defined by the half maximum of (3.7.7), we get the relation $\Delta\theta \approx 1/\gamma$.

An observer at a particular field point will receive signals emitted from a charged particle on a circular arc with length $\rho\Delta\theta \approx \rho/\gamma$ around a particular retarded position of it, because a signal emitted from a particular retarded point irradiates only within a small angle $\Delta\theta$. In other words, the radiation from a charged particle will irradiate a particular field point for the time interval measured at the retarded position of the charge,

$$\Delta t' = \frac{\rho}{\gamma v} \approx \frac{\rho}{\gamma c}. \tag{3.7.8}$$

Then the observer at the field point will see an electromagnetic pulse for the time duration,

$$\Delta t = \left\langle \frac{dt}{dt'} \right\rangle \Delta t' = \langle \eta \rangle \Delta t' \approx \frac{1}{\gamma^3} \frac{\rho}{c}, \tag{3.7.9}$$

where we used the relation $\langle \eta \rangle \approx 1/\gamma^2$. According to the Fourier analysis, this time duration of a pulse Δt corresponds to the frequency spread $\Delta\omega$,

$$\Delta\omega \approx \frac{2\pi}{\Delta t} \approx 2\pi\gamma^3 \frac{c}{\rho}. \tag{3.7.10}$$

c/ρ being the fundamental angular frequency corresponding to the circular motion of radius ρ. As seen from (3.7.10), a charged particle moving with extremely relativistic velocity radiates in a frequency spectrum which is so broad as to cover γ^3 times the fundamental angular frequency c/ρ.

The frequency spectrum of synchrotron radiation emitted per unit solid angle can be evaluated from (3.6.11) [3.1]. For simplicity, let a charged particle be moving around the time $t' = 0$ along a small curved orbit, which lies in the x-z plane and tangential to the z-axis at the origin at $t' = 0$ (see Fig. 3.5). Without a loss of generality, we can assume that the unit vector $\boldsymbol{i}_{t'}$ in (3.6.11) is in the y-z plane, making an angle θ with the z-axis. In addition, we define unit vectors $\boldsymbol{i}_{//} = \boldsymbol{i}_x$

and $\boldsymbol{i}_\perp$, which are perpendicular to each other and $\boldsymbol{i}_{t'}$. For an extremely relativistic particle, the radiation is significant only within a small angle $\theta \le 1/\gamma$. For this case, the unit vector $\boldsymbol{i}_\perp$ is very nearly perpendicular to the orbital plane. Furthermore, let an instantaneous radius of curvature for the orbit of a charged particle be denoted by ρ. Then, the vector amplitude part of the integrand in (3.6.11) can be expressed as

$$\boldsymbol{i}_{t'} \times \left(\boldsymbol{i}_{t'} \times \frac{\boldsymbol{v}}{c}\right) = \beta\left[-\boldsymbol{i}_{//} \sin\left(\frac{vt'}{\rho}\right) + \boldsymbol{i}_\perp \cos\left(\frac{vt'}{\rho}\right) \sin\theta\right]. \tag{3.7.11}$$

We note here that the radiation emitted by a charged particle is significant within a small angle $\theta \le 1/\gamma$ and the length of an orbit of the charge corresponding to it is much smaller than the instantaneous radius of curvature ρ. Hence Eq. (3.7.11) can be approximated as

$$\boldsymbol{i}_{t'} \times \left(\boldsymbol{i}_{t'} \times \frac{\boldsymbol{v}}{c}\right) \cong \beta\left[-\boldsymbol{i}_{//}\left(\frac{vt'}{\rho}\right) + \boldsymbol{i}_\perp \theta\right]. \tag{3.7.12}$$

Similarly, the phase part of the integrand in (3.6.11) can be rewritten as

$$\omega\left[t' - \boldsymbol{i}_{t'} \cdot \boldsymbol{r}_0(t')/c\right] = \omega\left[t' - \frac{\rho}{c} \sin\left(\frac{vt'}{\rho}\right) \cos\theta\right]. \tag{3.7.13}$$

For small values of vt'/ρ and θ, Eq. (3.7.13) can be approximated as

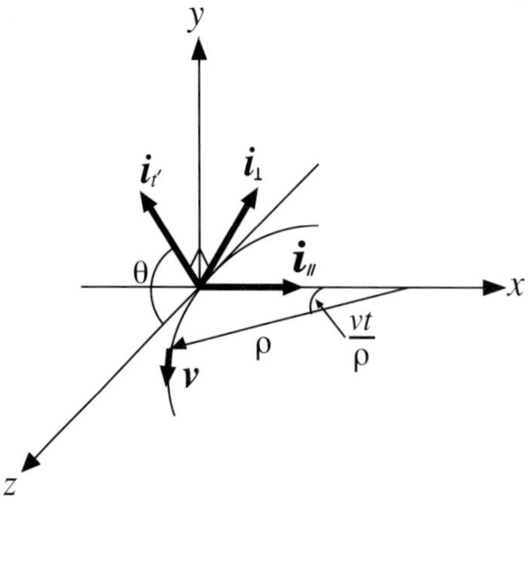

Fig. 3.5. Coordinate system for calculating the frequency spectrum of synchrotron radiation

$$\omega\left[t' - \boldsymbol{i}_{t'} \cdot \boldsymbol{r}_0(t')/c\right] \cong \frac{\omega}{2}\left[\left(\frac{1}{\gamma^2} + \theta^2\right)t' + \frac{1}{3}\left(\frac{c}{\rho}\right)^2 t'^3\right]. \tag{3.7.14}$$

Carrying out the integration (3.6.11) with the aid of the relations (3.7.12) and (3.7.14), we get for the frequency spectrum of synchrotron radiation emitted per unit solid angle,

$$\frac{d^2F(\omega,\theta)}{d\Omega d\omega} = \frac{3q^2}{4\pi^3}\sqrt{\frac{\mu_0}{\varepsilon_0}}\gamma^2\left(\frac{\omega}{\omega_{cr}}\right)^2\left[1+(\gamma\theta)^2\right]^2\left[K_{2/3}^2(\xi) + \frac{(\gamma\theta)^2}{1+(\gamma\theta)^2}K_{1/3}^2(\xi)\right], \tag{3.7.15}$$

where $K_{2/3}(\xi)$ and $K_{1/3}(\xi)$ denote the modified Bessel functions of the second kind, ξ being a parameter given by

$$\xi = \frac{\omega}{\omega_{cr}}\left[1+(\gamma\theta)^2\right]^{2/3}, \quad \omega_{cr} = 3\gamma^3\left(\frac{c}{\rho}\right). \tag{3.7.16}$$

In (3.7.15), the term proportional to $K_{2/3}^2(\xi)$ represents the radiation polarized in the plane of the orbit, while the term proportional to $K_{1/3}^2(\xi)$ corresponds to the radiation polarized in the plane perpendicular to the orbit. Integrating (3.7.15) over the whole solid angle, we obtain the frequency spectrum of the total energy emitted by synchrotron radiation,

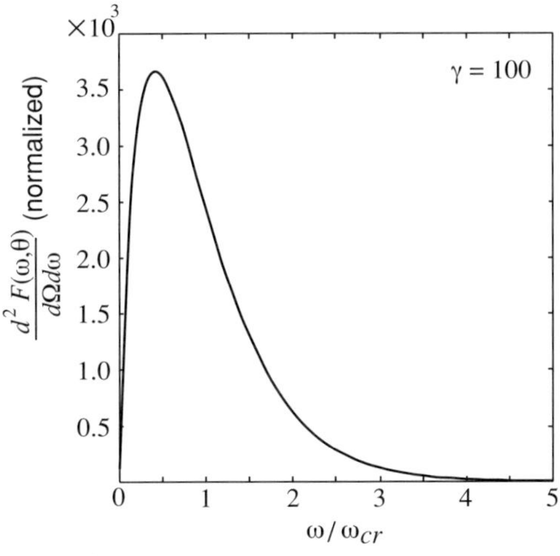

Fig. 3.6. Frequency spectrum of the power radiated per unit solid angle in the direction of $\theta = 0$

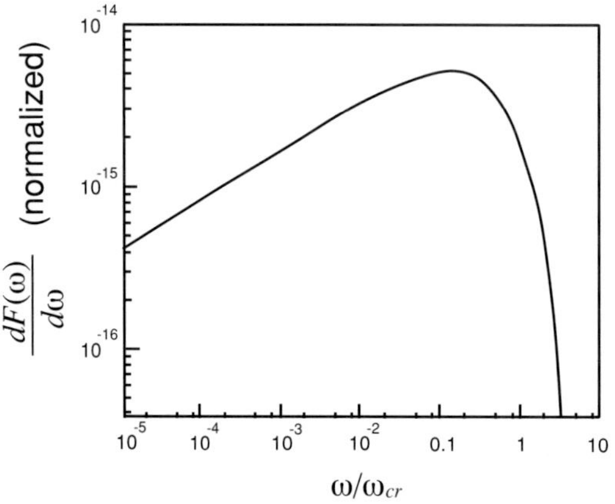

Fig. 3.7. Frequency spectrum of the total power emitted by synchrotron radiation

$$\frac{dF(\omega)}{d\omega} = \frac{\sqrt{3}q^2}{2\pi}\sqrt{\frac{\mu_0}{\varepsilon_0}}\gamma\left(\frac{\omega}{\omega_{cr}}\right)\int_{2\omega/\omega_{cr}}^{\infty} K_{5/3}(x)dx\,, \tag{3.7.17}$$

where $K_{5/3}(x)$ is the modified Bessel function of the second kind. For the lower extreme case of $\omega << \omega_{cr}$, Eq. (3.7.17) is found to vary as $(\omega\rho/c)^{1/3}$ within constant factors. On the other hand, for the upper extreme case of $\omega >> \omega_{cr}$, it behaves as $\gamma(\omega/\omega_{cr})^{1/2}\exp(-2\omega/\omega_{cr})$. In order to have a quantitative idea of the frequency spectrum for synchrotron radiation, we illustrate in Figs. 3.6 and 3.7 numerical examples of (3.7.15) for $\theta = 0$ and (3.7.17).

3.8 Cherenkov Radiation

In order for a charged particle to radiate electromagnetic waves, it must generally undergo an accelerated motion, as pointed out in Sect. 3.3. However, a charged particle moving uniformly in a material medium can also emit radiation if it is moving with a velocity greater than that of light in the medium. This radiation is possible only if the speed of light in the material medium is less than that in vacuum, or equiv-alently, the refractive index of the medium is greater than unity. The radiation due to a charged particle moving uniformly in a material medium was first observed by Cherenkov in 1934 [3.5]. After his name, this radiation is called the Cherenkov radiation. In 1937, Frank and Tamm presented a comprehensive theoretical interpretation of the Cherenkov radiation, clarifying all the peculiarities of the new phenomenon [3.6].

To have a qualitative idea of how the Cherenkov radiation occurs, let us refer to Figs. 3.8 and 3.9. Figure 3.8 shows how radiation is emitted by an accelerated charged particle which is linearly moving in vacuum with an arbitrary velocity or in a material medium with a velocity less than that of light in it. From the figure, we find for this case that radiated electromagnetic energy is concentrated in the forward direction or in the direction of motion of the charge. For this case, it should be noted that radiation never occurs unless a charged particle has acceleration. In Fig. 3.9, on the other hand, we illustrate the case where a charged particle is

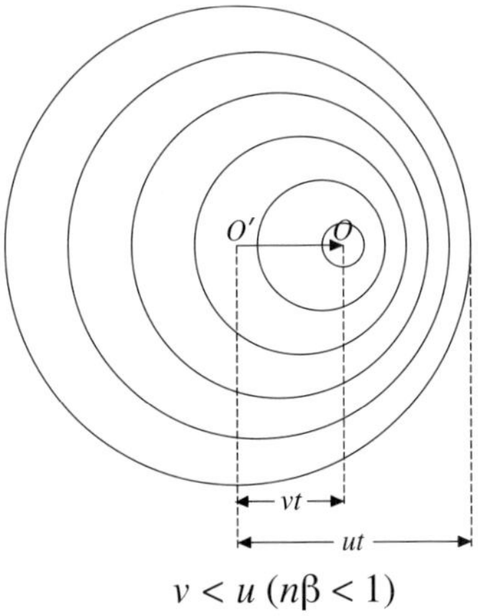

Fig. 3.8. Radiation from a charged particle moving with a velocity smaller than that of light in vacuum or in a material medium ($n\beta < 1$)

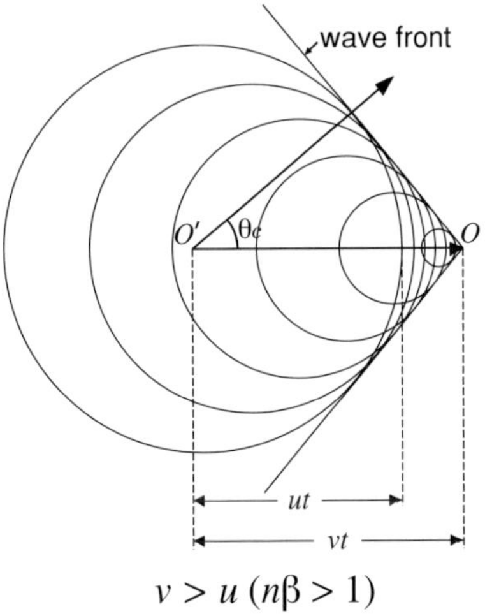

Fig. 3.9. Radiation from a charged particle moving with a velocity greater than that of light in a material medium ($n\beta > 1$)

moving uniformly in a material medium with a velocity greater than that of light in it. For the latter case, an electromagnetic disturbance emitted as a spherical wave at a particular point will be overtaken in the course of time by a charged particle which produced it. This is because the charge moves faster than the electromagnetic disturbance. Specifically, let the velocities of the charge and electromagnetic disturbance be v and u, respectively, and let the charge located at the point O' at the time 0 be transferred to the point O at the time t. Then, elementary spherical waves emitted between the times 0 and t are combined at the time t to form a shock wave with a conical wave front, which is propagated in the direction making an angle θ_c with the direction of motion of the charge. The angle θ_c is referred to as the Cherenkov angle. We note that the wave front of the elementary spherical wave emitted at the time 0 has moved by ut and the charge by vt when the time t has passed. Taking this fact into account and referring to Fig. 3.9, we find that the angle θ_c is given by

$$\cos\theta_c = \frac{u}{v} = \frac{1}{n\beta}, \tag{3.8.1}$$

where n denotes the refractive index of the material medium surrounding the charge, and β the ratio of the velocity of the charge to that of light in vacuum. From (3.8.1), we notice that we must have $n\beta > 1$ for the Cherenkov radiation to occur.

Let us consider the Cherenkov radiation in a quantitative manner [3.6–3.10]. For this purpose, let a charged particle be moving with constant velocity v along the z-axis in a material medium with a refractive index n. Then, the current density due to a uniformly moving charge is expressed as

$$\boldsymbol{J}(x, y, z, t) = \boldsymbol{i}_z qv\delta(x)\delta(y)\delta(z - vt), \tag{3.8.2}$$

which is rewritten in terms of the circular cylindrical coordinates,

$$\boldsymbol{J}(r, z, t) = \boldsymbol{i}_z qv\frac{\delta(r)}{2\pi r}\delta(z - vt). \tag{3.8.3}$$

In order to treat a general case of the Cherenkov radiation in a dispersive medium, we analyze it in frequency domain. Then, the Fourier component for the current density $\boldsymbol{J}$ is given by

$$\tilde{\boldsymbol{J}}(r, z, \omega) = \boldsymbol{i}_z\tilde{J}(r, z, \omega) = \boldsymbol{i}_z q\frac{\delta(r)}{2\pi r}e^{-j(\omega/v)z}. \tag{3.8.4}$$

From the expressions for the current density, we find that the vector potential $\boldsymbol{A}$ produced by it has only the z component. With the aid of (3.8.4), the Fourier component for the vector potential is found to satisfy the following inhomogeneous differential equation in frequency domain corresponding to (1.5.10):

$$\left[\frac{1}{r}\frac{\partial}{\partial r}r\frac{\partial}{\partial r}+\frac{\partial^2}{\partial z^2}+k^2\right]\tilde{A}_z(r,z,\omega)=-\mu\tilde{J}_z(r,z,\omega), \tag{3.8.5}$$

where $k = n\omega/c$, k denoting the wave number for the plane wave in a material medium with the refractive index n, which is assumed to be frequency dependent.

The solution of (3.8.5) can be represented in terms of an appropriately defined Green function $G(r,z;r',z',\omega)$, which satisfies an inhomogeneous differential equation

$$\left[\frac{1}{r}\frac{\partial}{\partial r}r\frac{\partial}{\partial r}+\frac{\partial^2}{\partial z^2}+k^2\right]G(r,z;r',z',\omega)=-\frac{1}{2\pi r}\delta(r-r')\delta(z-z'). \tag{3.8.6}$$

With the aid of (3.8.4), the solution of (3.8.5) can be expressed in terms of $G(r,z;r',z',\omega)$ as

$$\tilde{A}_z(r,z,\omega)=\mu q\int_0^{\infty}\int_{-\infty}^{\infty}G(r,z;r',z',\omega)\delta(r')e^{-j(\omega/v)z'}dr'dz'. \tag{3.8.7}$$

We first try to find the solution of (3.8.6) in the wave number domain. To this end, we represent $G(r,z;r',z',\omega)$ in terms of thc Fourier transform in the longitudinal wave number space as

$$G(r,z;r',z',\omega)=\frac{1}{2\pi}\int_{-\infty}^{\infty}g(r;r',\omega,k_z)e^{-jk_z(z-z')}dk_z. \tag{3.8.8}$$

Using (3.8.8) in (3.8.6), together with the Fourier transform for a delta function, we get a differential equation for $g(r;r',\omega,k_z)$ to satisfy,

$$\left[\frac{1}{r}\frac{d}{dr}r\frac{d}{dr}-k_r^2\right]g(r;r',\omega,k_z)=-\frac{1}{2\pi r}\delta(r-r'), \tag{3.8.9}$$

with

$$k_r^2=k_z^2-k^2. \tag{3.8.10}$$

The solution of (3.8.9) can be expressed in terms of the modified Bessel functions. In the regions $r > r'$ and $r < r'$, it takes the form

$$g(r;r',\omega,k_z)=C_1(r')K_0(k_r r) \quad (r>r'), \tag{3.8.11}$$

$$g(r;r',\omega,k_z)=C_2(r')I_0(k_r r) \quad (r<r'), \tag{3.8.12}$$

where I_0 and K_0 denote the modified Bessel functions of the first and second kinds, C_1 and C_2 being unknown constants to be determined from the boundary conditions at $r=r'$. First, the function g must be continuous at $r=r'$, from which we get

$$C_1(r')K_0(k_r r')=C_2(r')I_0(k_r r'). \tag{3.8.13}$$

On the other hand, its derivative is discontinuous at the same position. The discontinuity can be determined by integrating (3.8.9) around $r = r'$,

$$\left[r\frac{dg}{dr} \right]_{r=r'+0} - \left[r\frac{dg}{dr} \right]_{r=r'-0} = -\frac{1}{2\pi}. \tag{3.8.14}$$

Inserting (3.8.11) and (3.8.12) into (3.8.14) yields

$$k_r r' C_1(r') K_1(k_r r') + k_r r' C_2(r') I_1(k_r r') = \frac{1}{2\pi}. \tag{3.8.15}$$

With the aid of the identity concerning the modified Bessel functions

$$K_0(x)I_1(x) + K_1(x)I_0(x) = \frac{1}{x}, \tag{3.8.16}$$

we can determine unknowns C_1 and C_2. Thus we get for the solution of (3.8.9),

$$g(r; r', \omega, k_z) = \frac{1}{2\pi} I_0(k_r r) K_0(k_r r') \quad (0 \le r \le r'), \tag{3.8.17}$$

$$g(r; r', \omega, k_z) = \frac{1}{2\pi} K_0(k_r r) I_0(k_r r') \quad (0 \le r' \le r). \tag{3.8.18}$$

Now we can obtain, from (3.8.8) and (3.8.18), the Green function for the longitudinal component of the vector potential in frequency domain.

The longitudinal component of the vector potential produced by the current density (3.8.4) can be found by evaluating the integral (3.8.7) with the aid of (3.8.8), together with (3.8.18). After some manipulation, we get the final result for the axial component of the vector potential as follows:

$$\tilde{A}_z(r, z, \omega) = \frac{\mu q}{2\pi} K_0(k_v r) e^{-j(\omega / v)z}, \tag{3.8.19}$$

with

$$k_v = \frac{\omega}{v}\sqrt{1 - n^2\beta^2}. \tag{3.8.20}$$

In order to investigate the behavior of $\tilde{A}_z(r, z, \omega)$ at large radial distances, we note the asymptotic form of $K_0(k_v r)$,

$$K_0(k_v r) = \sqrt{\frac{\pi}{2k_v r}} e^{-k_v r} \quad (k_v r >> 1). \tag{3.8.21}$$

If $n\beta < 1$, k_v becomes real, and then the vector potential (3.8.19) represents an evanescent wave in the radial direction. On the other hand, if $n\beta > 1$, k_v becomes imaginary, which means that the vector potential (3.8.19) is a propagating wave in the radial direction. The latter case corresponds to the Cherenkov radiation. Hence the condition for the Cherenkov radiation to occur is given by

$$n\beta > 1 \text{ or } \quad v > \frac{c}{n}, \tag{3.8.22}$$

which coincides with the condition that the Cherenkov angle defined by (3.8.1) takes a real value.

The electric and magnetic fields corresponding to (3.8.19) are obtained from

$$\tilde{\boldsymbol{E}}(r,z,\omega) = -j\omega\tilde{\boldsymbol{A}}(r,z,\omega) - j\frac{1}{\omega\varepsilon\mu}\nabla\nabla\cdot\tilde{\boldsymbol{A}}(r,z,\omega), \tag{3.8.23}$$

$$\tilde{\boldsymbol{H}}(r,z,\omega) = \frac{1}{\mu}\nabla\times\tilde{\boldsymbol{A}}(r,z,\omega), \tag{3.8.24}$$

where $\tilde{\boldsymbol{A}}(r,z,\omega) = \boldsymbol{i}_z\tilde{A}_z(r,z,\omega)$, and the Lorentz condition in frequency domain corresponding to (1.5.9) was used. For $n\beta > 1$, we get, from (3.8.23) and (3.8.24), the following results for the electric and magnetic field components in frequency domain:

$$\tilde{E}_r(r,z,\omega) = j\frac{q|k_v|}{2\pi\varepsilon v}K_1\left(j|k_v|r\right)e^{-j(\omega/v)z}, \tag{3.8.25}$$

$$\tilde{E}_z(r,z,\omega) = -j\frac{q|k_v|}{2\pi\varepsilon v}\sqrt{n^2\beta^2-1}\,K_0\left(j|k_z|r\right)e^{-j(\omega/v)z}, \tag{3.8.26}$$

$$\tilde{H}_\varphi(r,z,\omega) = j\frac{q|k_v|}{2\pi}K_1\left(j|k_v|r\right)e^{-j(\omega/v)z}. \tag{3.8.27}$$

At large radial distances ($|k_v|r >> 1$), the field components in frequency domain reduce to

$$\tilde{E}_r(r,z,\omega) = \frac{q}{2\pi\varepsilon v}\sqrt{\frac{j\pi|k_v|}{2r}}\,e^{-j\left[|k_v|r+(\omega/v)z\right]}, \tag{3.8.28}$$

$$\tilde{E}_z(r,z,\omega) = -\frac{q}{2\pi\varepsilon v}\sqrt{\frac{j\pi|k_v|\left(n^2\beta^2-1\right)}{2r}}\,e^{-j\left[|k_v|r+(\omega/v)z\right]}, \tag{3.8.29}$$

$$\tilde{H}_\varphi(r,z,\omega) = \frac{q}{2\pi}\sqrt{\frac{j\pi|k_v|}{2r}}\,e^{-j\left[|k_v|r+(\omega/v)z\right]}. \tag{3.8.30}$$

The electromagnetic waves represented by (3.8.28) ~ (3.8.30) have locally the same properties as the plane waves. Specifically, as shown in Fig. 3.10, on the wave front, the electric and magnetic field vectors are perpendicular to each other and the wave vector $\boldsymbol{k}$, in the direction of which the waves propagate.

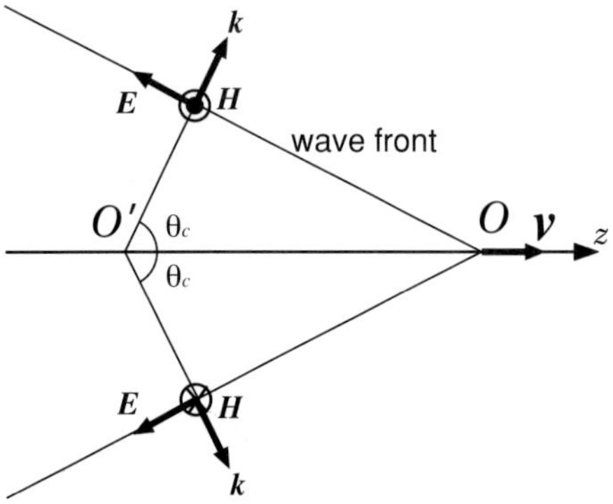

Fig. 3.10. Local properties of the fields produced by the Cherenkov radiation

Finally, let us try to find the frequency spectrum of the Cherenkov radiation, with the aid of (3.6.11). For this purpose, let a uniformly moving charged particle pass through a dielectric of finite thickness d during a time interval 2τ. For a uniform motion of the charge, its position vector is expressed as $\boldsymbol{r}_0(t') = \boldsymbol{i}_z vt'$, and let the direction of wave propagation $\boldsymbol{i}_{t'}$ make an angle θ with the direction of motion of the charge $\boldsymbol{i}_z$. Then, Eq. (3.6.11) reduces to

$$\frac{d^2F(\omega,\theta)}{d\omega d\Omega} = \frac{q^2\mu v^2}{4\pi u}\sin^2\theta\left|\frac{\omega}{2\pi}\int_{-\tau}^{\tau} e^{-j\omega t'(1-n\beta\cos\theta)}dt'\right|^2, \tag{3.8.31}$$

in which the integral on the right-hand side has a sharp maximum at the Cherenkov angle θ_c for $\omega\tau >> 1$, approaching a delta function as τ tends to infinity. Hence $\sin\theta$ in (3.8.31) can be replaced by $\sin\theta_c$ for $\omega\tau >>1$. The integral in (3.8.31) can be calculated as

$$\frac{\omega}{2\pi}\int_{-\tau}^{\tau} e^{-j\omega t'(1-n\beta\cos\theta)}dt' = \frac{\omega\tau}{\pi}\frac{\sin[\omega\tau(1-n\beta\cos\theta)]}{\omega\tau(1-n\beta\cos\theta)}. \tag{3.8.32}$$

The integration of (3.8.31) over the whole solid angle yields the energy radiated on a unit frequency interval at the frequency ω, or the frequency spectrum of the Cherenkov radiation,

$$\frac{dF(\omega)}{d\omega} = \frac{q^2\mu v^2}{4\pi u}\sin^2\theta_c\int_{-1}^{1} 2\pi\left(\frac{\omega\tau}{\pi}\right)^2 \frac{\sin^2[\omega\tau(1-n\beta\cos\theta)]}{[\omega\tau(1-n\beta\cos\theta)]^2}d(-\cos\theta)$$

$$= \frac{q^2\mu v\tau}{2\pi^2}\omega\sin^2\theta_c\int_{-\infty}^{\infty}\frac{\sin^2\xi}{\xi^2}d\xi$$

$$= \frac{q^2 \mu d}{4\pi} \omega \left(1 - \frac{1}{n^2 \beta^2}\right), \tag{3.8.33}$$

where the relation (3.8.1) for the Cherenkov angle was used. The approximation in the integration in the second line of (3.8.33) is valid for $\omega\tau >> 1$. From (3.8.33), it is confirmed that the Cherenkov radiation is possible only for those spectral regions for which $n\beta > 1$. Thus the total energy emitted by the Cherenkov radiation is obtained by integrating (3.8.33) over frequencies for which $n\beta > 1$,

$$U = \int_{n\beta>1} \frac{q^2 \mu d}{4\pi} \omega \left(1 - \frac{1}{n^2 \beta^2}\right) d\omega . \tag{3.8.34}$$

The relation (3.8.34) can also be obtained with the aid of the Fourier components for the electromagnetic fields produced by the Cherenkov radiation, which are given by (3.8.28) ~ (3.8.30). Let us evaluate the total energy radiated from a charged particle moving uniformly from $z = 0$ to $z = d$, neglecting the effects of refraction and diffraction. As seen from Fig. 3.11, the radiation for a particular frequency component emitted from a charged particle traveling the distance d propagates in the space between two cones with their apexes located at O' and O, with the corresponding Cherenkov angle θ_c. It should be noted here that the Cherenkov angle depends on frequency for dispersive media, as is evident from (3.8.1). Namely, different frequency component waves are emitted in different directions. Then, the total energy radiated for a particular frequency component can be calculated from the total Poynting energy passing through an annular surface S_0 perpendicular to the corresponding Cherenkov ray. This total Poynting energy is equal to the total energy passing through another annular surface S_1 normal to the z-

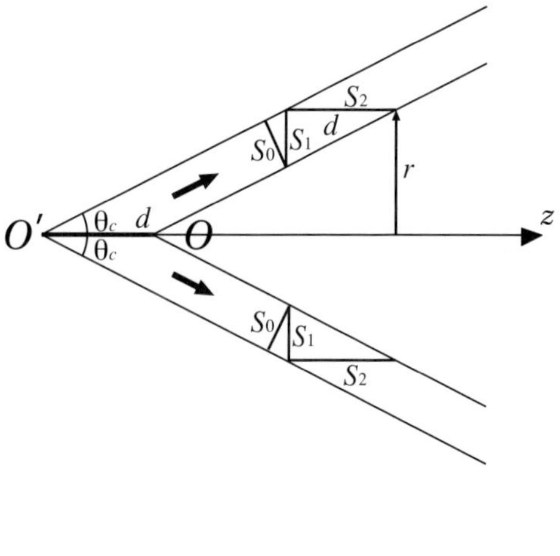

Fig. 3.11. Calculation of the total energy radiated

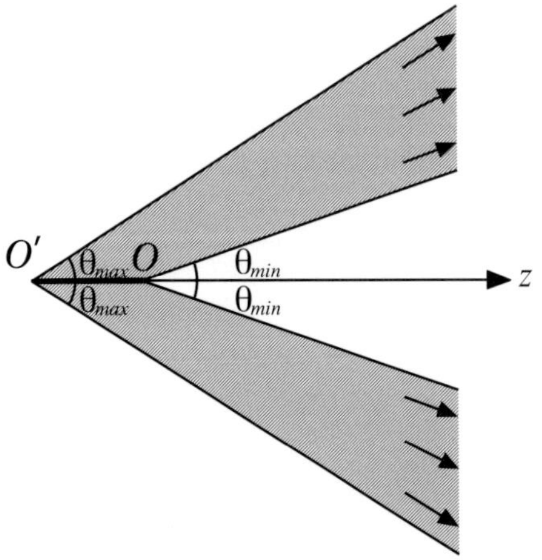

Fig. 3.12. Cherenkov radiation in a dispersive medium

axis or through a circular cylindrical surface S_2 of radius r and length d. Thus, summing up the contributions from all the frequency components for the radiated fields, we get for the overall total energy radiated from a charged particle traveling the distance d,

$$U = -2rd\int_0^\infty \tilde{E}_z(\omega)\tilde{H}_\varphi^*(\omega)d\omega\,, \tag{3.8.35}$$

which is found to be expressed also as

$$U = -2\pi rd\int_{-\infty}^\infty E_z(t)H_\varphi(t)dt\,, \tag{3.8.36}$$

according to our definition of the Fourier transforms (1.9.2) and (1.9.4). Then, inserting (3.8.29) and (3.8.30) in (3.8.35), we readily get (3.8.34).

As pointed out earlier, the Cherenkov radiation is possible only for the spectral regions for which $n\beta > 1$. Let the maximum and minimum values of n, which satisfy the condition $n\beta > 1$, be denoted by n_{max} and n_{min}, and the corresponding Cherenkov angles by θ_{max} and θ_{min}, which are defined as (3.8.1). Then, the Cherenkov radiation in a dispersive medium with the refractive index n is confined in the shaded region of space between the cone with its apex at O' and aperture $2\theta_{max}$ and the cone with its apex at O and aperture $2\theta_{min}$, as shown in Fig. 3.12.

In our analysis of the Cherenkov radiation, the velocity of a charged particle was assumed to be constant while it was radiating. However, as the charged particle emits radiation, it gradually slows down, ceasing to radiate any more after the condition $n\beta \le 1$ has been satisfied.

4. Macroscopic Theory of Relativistic Electron Beams

4.1 Modeling of Relativistic Electron Beams

An aggregate of electrons, which steadily flows with constant relativistic velocity in a particular direction, is called a relativistic electron beam. In this chapter, for simplicity, we treat a relativistic electron beam as a kind of continuous medium, or a kind of smooth fluid in macroscopic electromagnetic theory. Specifically, let us consider a macroscopically infinitesimal volume around a particular point in the electron beam, which yet contains a large number of electrons. Then, we assume that macroscopic quantities at this point such as the macroscopic electron velocity are defined as the average values of the corresponding microscopic quantities of electrons contained in the infinitesimal volume. Modeling an electron beam as a continuous medium specified above, we can investigate its electromagnetic properties from a macroscopic point of view.

Let a uniform aggregate of electrons with average number density N_0 be flowing with constant relativistic velocity $\boldsymbol{v}_0$ in the z direction. Then, in the electron beam, the electric field $\boldsymbol{E}_e$ and the magnetic-flux density $\boldsymbol{B}_e$ given by the following expressions are produced:

$$\boldsymbol{E}_e = -\boldsymbol{i}_r \frac{e}{2\varepsilon_0} N_0 r, \tag{4.1.1}$$

$$\boldsymbol{B}_e = -\boldsymbol{i}_\varphi \frac{e}{2\varepsilon_0 c^2} N_0 v_0 r, \tag{4.1.2}$$

where $-e$ denotes the electronic charge with the absolute value of 1.602×10^{-19}[C], and r is a radial distance measured from the axis of the electron beam. In addition, $\boldsymbol{i}_r$ and $\boldsymbol{i}_\varphi$ express unit vectors in the r and φ directions in the circular cylindrical coordinates, respectively. Due to the electric and magnetic fields given by (4.1.1) and (4.1.2), a particular electron at the position r in the electron beam is acted upon by the Lorentz force $\boldsymbol{F}$,

$$\boldsymbol{F} = -e\left(\boldsymbol{E}_e + \boldsymbol{v}_0 \times \boldsymbol{B}_e\right) = \boldsymbol{i}_r \frac{N_0 e^2}{2\varepsilon_0} r\left[1 - \left(\frac{v_0}{c}\right)^2\right]. \tag{4.1.3}$$

This Lorentz force is always directed outwardly in the radial direction. Due to this outward force, the electron beam spreads out in the radial direction, which prevents it from keeping a constant radius. In order to keep the electron beam from spreading out in the radial direction due to the self fields produced by it, and thus to guide it in a stable condition, we consider two typical methods in the subsequent discussion.

The first method to compensate for the static electric and magnetic fields given by (4.1.1) and (4.1.2) is to neutralize the space charge and steady current due to the electron beam by adding the flow of positive ions. For a special case where the static fields produced by the electron beam should be eliminated by positive ions with a valency of unity, we need a flow of positive ions with the same number density N_0 and the same drift velocity $\boldsymbol{v}_0$ as the electron beam. Experimentally, the ion-neutralized electron beam can be produced by injecting a pure electron beam into plasma or neutral gases.

The alternative method to remove the effect of the beam self-fields is to apply an intense magnetostatic field in the direction of beam flow and thus to suppress the transverse spread of electrons. As seen from the equation of motion for a charged particle moving in electromagnetic fields, (2.10.6), a sufficiently intense magnetostatic field imposed on the electron beam can keep the transverse velocities of electrons at a sufficiently small level, thus suppressing the transverse spread of the electron beam.

In order to treat the aforementioned two typical cases simultaneously, we assume that the electron beam is ion-neutralized and a finite magnetostatic field is applied in the direction of beam flow. In addition, it is assumed that the effects of the thermal motion of electrons and electron-ion collisions can be neglected. On the basis of a considerably simple model of the relativistic electron beam and the corresponding assumptions specified above, we can comprehend the essentials of its macroscopic properties.

4.2 Basic Field Equations for Small-Signal Fields

We make two assumptions to obtain the basic equations for describing oscillating electromagnetic fields that can exist in an ion-neutralized relativistic electron beam. First, we assume that oscillating fields in the electron beam set only electrons in a small oscillating motion, not affecting the motion of positive ions, because the mass of a positive ion is much larger than that of an electron. In addition, for a simplified treatment of the problem, we discuss only linear phenomena on the assumption of sufficiently small amplitudes of oscillating fields. This second assumption is referred to as small-signal approximations. Generally, we should note that electromagnetic fields in relativistic electron beams present complicated nonlinearities.

Let oscillating electromagnetic fields be turned on in a relativistic electron beam. Then, under small-signal approximations, electrons constituting the relativistic electron beam begin to oscillate with small amplitudes around their equilibrium positions, which yield oscillating charge and current distributions. Hence, a relativistic electron beam, as is the case with nonrelativistic electron beams, can be replaced by oscillating charges and currents distributed in vacuum from a viewpoint of electromagnetic theory for oscillating fields. It should be noted that these oscillating charges and currents depend upon the electromagnetic fields in the relativistic electron beam.

Let the average number density of electrons in the electron beam be N_0, and their oscillating number density n. Then, the density of space charges produced by the electron beam can be expressed as

$$\rho_e = -(N_0 + n)e = -N_0 e - ne. \tag{4.2.1}$$

The first term on the right-hand side of (4.2.1) is canceled out by the space charges due to positive ions, so the net charge density due to the electron beam, ρ is given by

$$\rho = -ne. \tag{4.2.2}$$

On the other hand, the current density produced by the motion of electrons constituting the electron beam, $\boldsymbol{J}_e$ can be represented in terms of the drift velocity of the electron beam, $\boldsymbol{v}_0$ and the oscillating velocity of the electron, $\boldsymbol{v}$ as

$$\begin{aligned} \boldsymbol{J}_e &= -(N_0 + n)e(\boldsymbol{v}_0 + \boldsymbol{v}) \\ &= -N_0 e\boldsymbol{v}_0 - N_0 e\boldsymbol{v} - ne\boldsymbol{v}_0 - ne\boldsymbol{v}, \end{aligned} \tag{4.2.3}$$

the first term of which is canceled out by the current density due to the flow of positive ions, and the fourth term of which is of second-order with respect to oscillating fields, being neglected under small-signal approximations. Hence, the small-signal current density produced by the oscillating motion of electrons, $\boldsymbol{J}$ reduces to

$$\boldsymbol{J} = -N_0 e\boldsymbol{v} - ne\boldsymbol{v}_0, \tag{4.2.4}$$

which can be rewritten, with the aid of (4.2.2), as

$$\boldsymbol{J} = -N_0 e\boldsymbol{v} + \rho\boldsymbol{v}_0, \tag{4.2.5}$$

the first term of which represents the current density component due to a small variation in the electron velocity, and the second term of which corresponds to the current density component due to a small variation in the number density of electrons.

As mentioned above, a relativistic electron beam can be replaced by oscillating charges and currents distributed in vacuum, from a viewpoint of electromagnetic theory for oscillating fields. Hence the oscillating electromagnetic fields in a relativistic electron beam can be related to the charge density (4.2.2) and the current density (4.2.4) or (4.2.5) through the Maxwell equations in vacuum,

$$\nabla \times \boldsymbol{E} = -\frac{\partial \boldsymbol{B}}{\partial t}, \tag{4.2.6}$$

$$\nabla \times \frac{\boldsymbol{B}}{\mu_0} = \frac{\partial}{\partial t}\varepsilon_0 \boldsymbol{E} + \boldsymbol{J}, \tag{4.2.7}$$

$$\nabla \cdot \varepsilon_0 \boldsymbol{E} = \rho, \tag{4.2.8}$$

$$\nabla \cdot \boldsymbol{B} = 0, \tag{4.2.9}$$

where the charge and current densities ρ and $\boldsymbol{J}$ are not independent sources but depend upon the fields $\boldsymbol{E}$ and $\boldsymbol{B}$ in the electron beam, as mentioned in the foregoing discussion. In addition, for the Maxwell equations (4.2.6) ~ (4.2.9) to have unique solutions, the unknown $\boldsymbol{v}$ included in $\boldsymbol{J}$ in (4.2.4) must be represented in terms of the fields $\boldsymbol{E}$ and $\boldsymbol{B}$.

The relation of the oscillating velocity $\boldsymbol{v}$ to the fields $\boldsymbol{E}$ and $\boldsymbol{B}$ can be obtained from the relativistic equation of motion for the electron, (2.10.6), being given by

$$\frac{d\boldsymbol{v}}{dt} = -\frac{e}{\gamma m_0}\left[\boldsymbol{E} - \frac{\boldsymbol{v}_0}{c^2}\left(\boldsymbol{v}_0 \cdot \boldsymbol{E}\right) + \boldsymbol{v}_0 \times \boldsymbol{B} + \boldsymbol{v} \times \boldsymbol{B}_0\right], \tag{4.2.10}$$

with

$$\gamma = \frac{1}{\sqrt{1-\beta_0^2}}, \quad \beta_0 = \frac{v_0}{c}, \tag{4.2.11}$$

where m_0 denotes the rest mass of the electron, being equal to 9.109×10^{-31}[kg], and $\boldsymbol{B}_0$ the magnetic-flux density for a finite magnetostatic field applied in the direction of beam flow. The temporal derivative on the left-hand side of (4.2.10) means the differentiation with respect to time that takes into account the translational motion of the electron, namely,

$$\frac{d\boldsymbol{v}}{dt} = \frac{\partial \boldsymbol{v}}{\partial t} + (\boldsymbol{v}_0 \cdot \nabla)\boldsymbol{v}. \tag{4.2.12}$$

In deriving (4.2.10) from (2.10.6), we assumed that $|\boldsymbol{v}| << |\boldsymbol{v}_0|$, neglecting terms of orders higher than the first with respect to $\boldsymbol{v}$.

With the aid of (4.2.2), (4.2.5), and (4.2.10), the Maxwell equations (4.2.6) ~ (4.2.9) can be uniquely solved to determine the electromagnetic fields in a relativistic electron beam. Incidentally, it should be noted that instead of (4.2.8) we can use the equation of continuity

$$\nabla \cdot \boldsymbol{J} = -\frac{\partial \rho}{\partial t}, \tag{4.2.13}$$

to find the electromagnetic fields in a relativistic electron beam.

In the above discussion, electron beams were replaced by charges and currents distributed in vacuum from a viewpoint of electromagnetic theory. Alternatively, we can discuss the interaction of electromagnetic fields with electron beams macroscopically by the method equivalent to the former method. In the latter method, we treat electron beams as moving dispersive media. In the subsequent discussion, we will find the basic equations for describing electromagnetic fields in a relativistic electron beam on the basis of the latter method. The motion of the electron described by (4.2.10) is a small oscillation superimposed on a uniform translational motion. From an alternative point of view, the electron in such a motion generally produces not only an electric dipole moment but also a magnetic dipole moment. It is well known that an electron in a simple harmonic oscillation produces an electric dipole moment. Then, if we watch such an electron from an inertial system moving with constant velocity relative to the electron, we will see a magnetic dipole moment in addition to an electric dipole moment.

Let a particular electron in the electron beam be displaced, under the influence of oscillating fields, by a small distance $\boldsymbol{s}$ from an equilibrium position it would occupy in the absence of oscillating fields. Then, the densities of electric and magnetic dipole moments produced by an aggregate of electrons in the motion specified by (4.2.10), $\boldsymbol{P}$ and $\boldsymbol{M}$ are expressed as

$$\boldsymbol{P} = -N_0 e \boldsymbol{s}, \tag{4.2.14}$$

$$\boldsymbol{M} = -\boldsymbol{v}_0 \times \boldsymbol{P}. \tag{4.2.15}$$

In view of the relation $\boldsymbol{v} = d\boldsymbol{s}/dt$, a small displacement $\boldsymbol{s}$ is found, from (4.2.10), to satisfy the following relation:

$$\frac{d^2\boldsymbol{s}}{dt^2} = -\frac{e}{\gamma m_0}\left[\boldsymbol{E} - \frac{\boldsymbol{v}_0}{c^2}(\boldsymbol{v}_0 \cdot \boldsymbol{E}) + \boldsymbol{v}_0 \times \boldsymbol{B} + \frac{d\boldsymbol{s}}{dt} \times \boldsymbol{B}_0\right]. \tag{4.2.16}$$

Equation (4.2.14) is a well-known definition of the density of electric dipole moments. On the other hand, Eq. (4.2.15) represents the density of magnetic dipole

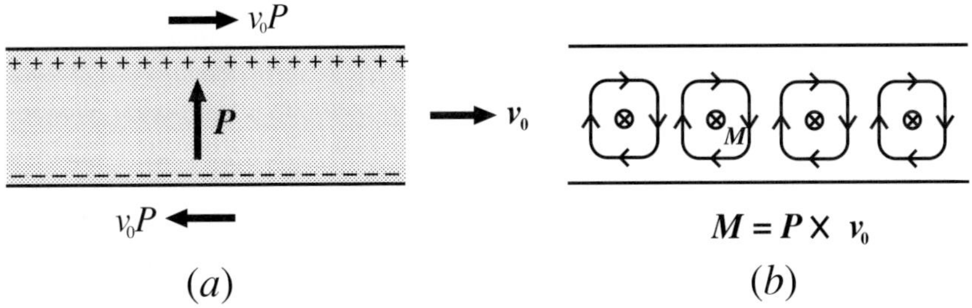

Fig. 4.1. A moving polarized dielectric slab

moments produced by the motion of electric dipoles. The reason why the magnetic dipole moment corresponding to (4.2.15) is produced by the motion of electric dipoles can be realized by referring to a polarized dielectric slab in translational motion as shown in Fig. 4.1. On the upper and lower surfaces of a moving polarized dielectric slab, there appear surface currents flowing in opposite directions to each other. These surface currents are equivalent to a chain of loop currents flowing in the interior of the dielectric slab, which produce magnetic dipole moments with the density given by (4.2.15).

With the aid of the densities of electric and magnetic dipole moments, $\boldsymbol{P}$ and $\boldsymbol{M}$, defined by (4.2.14) and (4.2.15), the charge and current densities ρ and $\boldsymbol{J}$ given by (4.2.2) and (4.2.5) are rewritten as

$$\rho = -\nabla \cdot \boldsymbol{P}, \tag{4.2.17}$$

$$\boldsymbol{J} = \frac{\partial \boldsymbol{P}}{\partial t} + \nabla \times \boldsymbol{M}, \tag{4.2.18}$$

where the relation $\boldsymbol{v} = d\boldsymbol{s}/dt$ is used in deriving (4.2.18). Inserting (4.2.17) in (4.2.8) and (4.2.18) in (4.2.7), and then defining new field vectors $\boldsymbol{D}$ and $\boldsymbol{H}$ in the resultant equations as

$$\boldsymbol{D} = \varepsilon_0 \boldsymbol{E} + \boldsymbol{P}, \quad \boldsymbol{H} = \frac{\boldsymbol{B}}{\mu_0} - \boldsymbol{M}, \tag{4.2.19}$$

we can rewrite the Maxwell equations (4.2.6) ~ (4.2.9) as

$$\nabla \times \boldsymbol{E} = -\frac{\partial \boldsymbol{B}}{\partial t}, \tag{4.2.20}$$

$$\nabla \times \boldsymbol{H} = \frac{\partial \boldsymbol{D}}{\partial t}, \tag{4.2.21}$$

$$\nabla \cdot \boldsymbol{D} = 0, \tag{4.3.22}$$

$$\nabla \cdot \boldsymbol{B} = 0, \tag{4.2.23}$$

which are found to take the same form as the Maxwell equations in material media given by (1.2.8) ~ (1.2.11).

In this section, we presented two equivalent methods to describe electromagnetic fields in a relativistic electron beam. The first method is called the convection current model, in which the electron beam is modeled by the convection current it carries and the accompanying charge. On the other hand, the second method is referred to as the polarization current model, in which the electron beam is represented by the properly defined polarization current and the accompanying magnetization current.

4.3 Constitutive Relations for Small-Signal Fields

4.3.1 Constitutive Relation in the Convection Current Model

If the Maxwell equations in the convection current model, (4.2.6) ~ (4.2.9) are to be solved uniquely, the current density (4.2.4) or (4.2.5) must be represented in terms of the field vectors $\boldsymbol{E}$ and $\boldsymbol{B}$. In order to simplify the analysis, let us represent field vectors in terms of the Fourier integral as

$$\boldsymbol{E}(\boldsymbol{r},t) = \frac{1}{(2\pi)^2}\iint \tilde{\boldsymbol{E}}(\boldsymbol{r}_\perp,\omega,k_z)e^{j(\omega t - k_z z)}d\omega dk_z, \tag{4.3.1}$$

where $\tilde{\boldsymbol{E}}$ is assumed to be a function of transverse coordinates, angular frequency ω, and wave number in the z direction, k_z. Then, noting that the temporal derivative operator defined in (4.2.12) is represented in complex form as

$$\frac{d}{dt} = j(\omega - v_0 k_z), \tag{4.3.2}$$

the equation of motion for the electron, (4.2.10) can be rewritten in component form as

$$\hat{A}\cdot\tilde{\boldsymbol{v}} = -\frac{e}{\gamma m_0}\left[\tilde{\boldsymbol{E}} - \frac{\boldsymbol{v}_0}{c^2}(\boldsymbol{v}_0\cdot\tilde{\boldsymbol{E}}) + \boldsymbol{v}_0\times\tilde{\boldsymbol{B}}\right], \tag{4.3.3}$$

with

$$\hat{A} = \begin{bmatrix} j(\omega - v_0 k_z) & \omega_c & 0 \\ -\omega_c & j(\omega - v_0 k_z) & 0 \\ 0 & 0 & j(\omega - v_0 k_z) \end{bmatrix}, \tag{4.3.4}$$

$$\omega_c = \frac{eB_0}{\gamma m_0}, \tag{4.3.5}$$

where ω_c denotes the cyclotron angular frequency for the electron in the laboratory frame, which is an inertial system to watch a relativistic electron beam moving with drift velocity $\boldsymbol{v}_0$.

Solving (4.3.3) for $\tilde{\boldsymbol{v}}$, we get

$$\tilde{\boldsymbol{v}} = \hat{T}\cdot\left[\tilde{\boldsymbol{E}} - \frac{\boldsymbol{v}_0}{c^2}(\boldsymbol{v}_0\cdot\tilde{\boldsymbol{E}}) + \boldsymbol{v}_0\times\tilde{\boldsymbol{B}}\right], \tag{4.3.6}$$

with

$$\hat{T} = \frac{e}{\gamma m_0}\hat{A}^{-1} = \begin{bmatrix} T_{11} & T_{12} & 0 \\ -T_{12} & T_{11} & 0 \\ 0 & 0 & T_{33} \end{bmatrix}, \tag{4.3.7}$$

$$T_{11} = j\frac{e}{\gamma m_0}\frac{\omega - v_0 k_z}{\left(\omega - v_0 k_z\right)^2 - \omega_c^2}, \tag{4.3.8}$$

$$T_{12} = -\frac{e}{\gamma m_0}\frac{\omega_c}{\left(\omega - v_0 k_z\right)^2 - \omega_c^2}, \tag{4.3.9}$$

$$T_{33} = j\frac{e}{\gamma m_0}\frac{1}{\omega - v_0 k_z}, \tag{4.3.10}$$

where $\hat{A}^{-1}$ denotes the inverse of $\hat{A}$.

Substituting (4.3.6) in (4.2.5), we find that the current density $\tilde{\boldsymbol{J}}$ can be expressed in terms of the field vectors $\tilde{\boldsymbol{E}}$ and $\tilde{\boldsymbol{B}}$ as

$$\tilde{\boldsymbol{J}} - \tilde{\rho}\boldsymbol{v}_0 = \gamma\hat{\sigma}\cdot\left[\tilde{\boldsymbol{E}} - \frac{\boldsymbol{v}_0}{c^2}(\boldsymbol{v}_0\cdot\tilde{\boldsymbol{E}}) + \boldsymbol{v}_0\times\tilde{\boldsymbol{B}}\right], \tag{4.3.11}$$

with

$$\hat{\sigma} = \begin{bmatrix} \sigma_{11} & \sigma_{12} & 0 \\ -\sigma_{12} & \sigma_{11} & 0 \\ 0 & 0 & \sigma_{33} \end{bmatrix}, \tag{4.3.12}$$

$$\sigma_{11} = -j\frac{\omega'\varepsilon_0 X}{1-Y^2}, \quad \sigma_{12} = \frac{\omega'\varepsilon_0 XY}{1-Y^2}, \quad \sigma_{33} = -j\omega'\varepsilon_0 X, \tag{4.3.13}$$

$$X = \left(\frac{\omega_p}{\omega'}\right)^2, \quad Y = \frac{\omega_c'}{\omega'}, \tag{4.3.14}$$

$$\omega' = \gamma(\omega - v_0 k_z), \tag{4.3.15}$$

$$\omega_p = \sqrt{\frac{N_0 e^2}{\gamma m_0 \varepsilon_0}}, \quad \omega_c' = \gamma\omega_c. \tag{4.3.16}$$

In the above equations, ω_p denotes the plasma angular frequency for the electron beam, which is an invariant independent of the drift velocity of the electron beam. In addition, ω' and ω_c' are the angular frequency of the electromagnetic wave and the cyclotron angular frequency for the electron in the rest frame of the electron beam. The tensor $\hat{\sigma}$ expresses the conductivity tensor for the electron beam in its rest frame.

Equation (4.3.11) is the constitutive relation for small-signal fields in the convection current model, which we find has the same form as the constitutive relation for a moving conducting medium given by (2.4.8). However, it should be noted that the constitutive relation (4.3.11) is different from (2.4.8) in the following two respects. First, the conductivity in (2.4.8) is a scalar, while that in (4.3.11) becomes a tensor due to the effect of a magnetostatic field applied on the electron beam. Second, in (2.4.8), the part of the current dependent on the conductivity corresponds to a dissipative current converted to the Joule heat. On the other hand, the current in (4.3.11) is never converted to the Joule heat, because we have neglected the effect of electron-ion collisions, which is the cause of dissipation of energy in the electron beam.

4.3.2 Constitutive Relations in the Polarization Current Model

The basic field equations in the polarization current model are given by (4.2.20) ~ (4.2.23). For these equations to be solved uniquely, the field vectors $\tilde{\boldsymbol{D}}$ and $\tilde{\boldsymbol{H}}$ must be represented in terms of the field vectors $\tilde{\boldsymbol{E}}$ and $\tilde{\boldsymbol{B}}$ with the aid of (4.2.19). For this purpose, in view of (4.3.2), we rewrite the equation of motion for the electron, (4.2.16) in the following form:

$$\hat{F} \cdot \tilde{s} = -\frac{e}{\gamma m_0} \left[\tilde{\boldsymbol{E}} - \frac{\boldsymbol{v}_0}{c^2} (\boldsymbol{v}_0 \cdot \tilde{\boldsymbol{E}}) + \boldsymbol{v}_0 \times \tilde{\boldsymbol{B}} \right], \tag{4.3.17}$$

where

$$\hat{F} = \begin{bmatrix} -(\omega - v_0 k_z)^2 & j\omega_c(\omega - v_0 k_z) & 0 \\ -j\omega_c(\omega - v_0 k_z) & -(\omega - v_0 k_z)^2 & 0 \\ 0 & 0 & -(\omega - v_0 k_z)^2 \end{bmatrix}. \tag{4.3.18}$$

Solving (4.3.17) for $\tilde{s}$, and using it in (4.2.14), we get for the density of electric dipole moments, or the polarization vector $\tilde{\boldsymbol{P}}$, produced by the electron beam,

$$\tilde{\boldsymbol{P}} = \gamma^2 \varepsilon_0 (\hat{\varepsilon} - \hat{u}) \cdot \left[\tilde{\boldsymbol{E}} - \frac{\boldsymbol{v}_0}{c^2} (\boldsymbol{v}_0 \cdot \tilde{\boldsymbol{E}}) + \boldsymbol{v}_0 \times \tilde{\boldsymbol{B}} \right], \tag{4.3.19}$$

where $\hat{u}$ is a unit tensor, and $\hat{\varepsilon}$ denotes an equivalent relative permittivity tensor for the electron beam in its rest frame defined by

$$\hat{\varepsilon} = \begin{bmatrix} \varepsilon_{11} & j\varepsilon_{12} & 0 \\ -j\varepsilon_{12} & \varepsilon_{11} & 0 \\ 0 & 0 & \varepsilon_{33} \end{bmatrix}, \tag{4.3.20}$$

with

$$\varepsilon_{11} = 1 - \frac{X}{1 - Y^2}, \quad \varepsilon_{12} = -\frac{XY}{1 - Y^2}, \quad \varepsilon_{33} = 1 - X, \tag{4.3.21}$$

where X and Y are given by (4.3.14). The equivalent relative permittivity tensor for the electron beam in its rest frame, $\hat{\varepsilon}$ is found to take exactly the same form as that for a stationary magnetized electron plasma. Thus, from a viewpoint of electromagnetic theory, the electron beam is equivalent to an electron plasma moving with constant velocity.

Each element of the equivalent relative permittivity tensor for the electron beam given by (4.3.20) depends on the frequency of the electromagnetic wave in its rest frame. Thus, in its rest frame, the electron beam can be regarded as a dielectric with frequency dispersion. On the other hand, viewed from the laboratory frame, each element of the equivalent permittivity tensor for it becomes a function of both frequency and wave number. Hence, in the laboratory frame, it behaves like a dielectric with both frequency and wave number dispersion.

Let us insert the polarization $\tilde{\boldsymbol{P}}$ given by (4.3.19) and the magnetization $\tilde{\boldsymbol{M}}$ defined by (4.2.15) in (4.2.19). Then, after some manipulation, we get for the constitutive relations for small-signal fields in the polarization current model for a relativistic electron beam,

$$\tilde{\boldsymbol{D}} = \hat{\xi} \cdot \tilde{\boldsymbol{E}} + \hat{\eta} \cdot \tilde{\boldsymbol{B}}, \tag{4.3.22}$$

$$\tilde{\boldsymbol{H}} = \hat{\eta} \cdot \tilde{\boldsymbol{E}} + \hat{\chi} \cdot \tilde{\boldsymbol{B}}, \tag{4.3.23}$$

where $\hat{\xi}$, $\hat{\eta}$, and $\hat{\chi}$ are tensors defined by

$$\hat{\xi} = \gamma^2 \varepsilon_0 \begin{bmatrix} \varepsilon_{11} - \beta_0^2 & j\varepsilon_{12} & 0 \\ -j\varepsilon_{12} & \varepsilon_{11} - \beta_0^2 & 0 \\ 0 & 0 & \varepsilon_{33}/\gamma^2 \end{bmatrix}, \tag{4.3.24}$$

$$\hat{\eta} = \gamma^2 \sqrt{\frac{\varepsilon_0}{\mu_0}} \begin{bmatrix} j\beta_0\varepsilon_{12} & -\beta_0(\varepsilon_{11} - 1) & 0 \\ \beta_0(\varepsilon_{11} - 1) & j\beta_0\varepsilon_{12} & 0 \\ 0 & 0 & 0 \end{bmatrix}, \tag{4.3.25}$$

$$\hat{\chi} = \gamma^2 \frac{1}{\mu_0} \begin{bmatrix} 1 - \beta_0^2\varepsilon_{11} & -j\beta_0^2\varepsilon_{12} & 0 \\ j\beta_0^2\varepsilon_{12} & 1 - \beta_0^2\varepsilon_{11} & 0 \\ 0 & 0 & 1/\gamma^2 \end{bmatrix}. \tag{4.3.26}$$

As seen from (4.3.22) and (4.3.23), in the constitutive relations for a relativistic electron beam in the polarization current model, the field vectors $\tilde{\boldsymbol{D}}$ and $\tilde{\boldsymbol{H}}$ depend on both the field vectors $\tilde{\boldsymbol{E}}$ and $\tilde{\boldsymbol{B}}$. In addition, the material constants for a relativistic electron beam are given by the tensors (4.3.24) ~ (4.3.26), each element of which is a function of both frequency and wave number. Hence, from a viewpoint of macroscopic electromagnetic theory, a relativistic electron beam can be regarded as a bianisotropic medium with frequency and wave number dispersion.

The material constants for a relativistic electron beam, $\hat{\xi}$, $\hat{\eta}$, and $\hat{\chi}$ are found to have the following symmetry properties:

$$\hat{\xi} = \hat{\xi}^{\dagger}, \quad \hat{\chi} = \hat{\chi}^{\dagger}, \quad \hat{\eta} = -\hat{\eta}^{\dagger}, \tag{4.3.27}$$

where the sign † attached to the tensors for material constants means the Hermitian conjugates of the functions, or the complex conjugates of their transpose tensors. The above conditions represent that a relativistic electron beam as a dispersive bianisotropic medium is loss-free.

The constitutive relations for a relativistic electron beam in the polarization current model, (4.3.22) and (4.3.23) are rewritten also as

$$\tilde{\boldsymbol{D}} = \hat{\kappa} \cdot \tilde{\boldsymbol{E}} + \hat{\tau} \cdot \tilde{\boldsymbol{H}}, \tag{4.3.28}$$

$$\tilde{\boldsymbol{B}} = \hat{\tau} \cdot \tilde{\boldsymbol{E}} + \hat{\mu} \cdot \tilde{\boldsymbol{H}}, \tag{4.3.29}$$

where

$$\hat{\kappa} = \hat{\xi} - \hat{\eta}\hat{\chi}^{-1}\hat{\eta}, \quad \hat{\tau} = \hat{\chi}^{-1}\hat{\eta}, \quad \hat{\mu} = \hat{\chi}^{-1}. \tag{4.3.30}$$

With the aid of (4.3.24) ~ (4.3.26), the tensors for material constants $\hat{\kappa}$, $\hat{\tau}$, and $\hat{\mu}$ are found to take the form

$$\hat{\kappa} = \varepsilon_0 \begin{bmatrix} K_1 & jK_2 & 0 \\ -jK_2 & K_1 & 0 \\ 0 & 0 & \varepsilon_{33} \end{bmatrix}, \tag{4.3.31}$$

$$\hat{\tau} = \frac{1}{c} \begin{bmatrix} jL_1 & -L_2 & 0 \\ L_2 & jL_1 & 0 \\ 0 & 0 & 0 \end{bmatrix}, \tag{4.3.32}$$

$$\hat{\mu} = \mu_0 \begin{bmatrix} M_1 & jM_2 & 0 \\ -jM_2 & M_1 & 0 \\ 0 & 0 & 1 \end{bmatrix}, \tag{4.3.33}$$

with

$$K_1 = \frac{\varepsilon_{11} - \beta_0^2\left(\varepsilon_{11}^2 - \varepsilon_{12}^2\right)}{\gamma^2 \Delta}, \quad K_2 = \frac{\varepsilon_{12}}{\gamma^2 \Delta}, \tag{4.3.34}$$

$$M_1 = \frac{1 - \beta_0^2 \varepsilon_{11}}{\gamma^2 \Delta}, \quad M_2 = \frac{\beta_0^2 \varepsilon_{12}}{\gamma^2 \Delta}, \tag{4.3.35}$$

$$L_1 = \frac{\beta_0 \varepsilon_{12}}{\gamma^2 \Delta}, \quad L_2 = \frac{\beta_0}{\Delta}\left[(\varepsilon_{11} - 1)(1 - \beta_0^2 \varepsilon_{11}) + \beta_0^2 \varepsilon_{12}^2\right], \tag{4.3.36}$$

$$\Delta = (1-\beta_0^2\varepsilon_{11})^2 - \beta_0^4\varepsilon_{12}^2. \tag{4.3.37}$$

If the material medium whose constitutive relations are given by (4.3.28) and (4.3.29) is to be loss-free, the material constants $\hat{\kappa}$, $\hat{\mu}$, and $\hat{\tau}$ must satisfy the following conditions:

$$\hat{\kappa} = \hat{\kappa}^\dagger, \quad \hat{\mu} = \hat{\mu}^\dagger, \quad \hat{\tau} = -\hat{\tau}^\dagger, \tag{4.3.38}$$

which can be confirmed with the aid of (4.3.31) ~ (4.3.33).

4.4 Boundary Conditions at the Beam Boundary

Let us consider the boundary conditions that the field vectors satisfy at the boundary between the electron beam and the surrounding medium. For either of the convection current model and the polarization current model, the boundary conditions can be obtained from the integral representations for the basic field equations, namely, (4.2.6) ~ (4.2.9) for the former model and (4.2.20) ~ (4.2.23) for the latter model. The procedure for working out the boundary conditions is the same as presented in Sect. 1.4.

4.4.1 Boundary Conditions in the Convection Current Model

Applying the integral representations for the basic field equations for this model, (4.2.6) ~ (4.2.9) to a small domain around a point on the beam boundary with appropriate limiting processes, we find the following boundary conditions for this model:

$$\boldsymbol{n} \times (\boldsymbol{E}_1 - \boldsymbol{E}_2) = 0, \tag{4.4.1}$$

$$\boldsymbol{n} \times (\boldsymbol{B}_1 - \boldsymbol{B}_2) = \mu_0 \boldsymbol{K}, \tag{4.4.2}$$

$$\boldsymbol{n} \cdot (\varepsilon_0 \boldsymbol{E}_1 - \varepsilon_0 \boldsymbol{E}_2) = \xi, \tag{4.4.3}$$

$$\boldsymbol{n} \cdot (\boldsymbol{B}_1 - \boldsymbol{B}_2) = 0, \tag{4.4.4}$$

where subscripts 1 and 2 refer to the regions outside and inside the beam, respectively, and $\boldsymbol{n}$ denotes a unit normal vector directed from region 2 to region 1. In addition, the region outside the beam is assumed to be vacuum. Furthermore, $\boldsymbol{K}$ and ξ express the densities of surface current and charge appearing on the beam surface.

Noting that the surface current on the beam surface is produced by the surface charge moving with the drift velocity of the beam, $\boldsymbol{v}_0$, we have the following relation:

$$\boldsymbol{K} = \xi \boldsymbol{v}_0. \tag{4.4.5}$$

Combining (4.4.2) and (4.4.3), with the aid of (4.4.5), we get

$$\boldsymbol{n} \times (\boldsymbol{B}_1 - \boldsymbol{B}_2) = \frac{1}{c^2}\left[\boldsymbol{n} \cdot (\boldsymbol{E}_1 - \boldsymbol{E}_2)\right]\boldsymbol{v}_0 . \quad (4.4.6)$$

Eequations (4.4.1), (4.4.4), and (4.4.6) represent the boundary conditions that the field vectors $\boldsymbol{E}$ and $\boldsymbol{B}$ satisfy at the boundary between the beam and the surrounding vacuum region, which replace the conditions (4.4.1) ~ (4.4.4). It should be noted that the condition (4.4.4) is automatically satisfied if the conditions (4.4.1) and (4.4.6) are met.

4.4.2 Boundary Conditions in the Polarization Current Model

Starting from the integral representations for the basic field equations (4.2.20) ~ (4.2.23) for the polarization current model, and following the same steps as for the convection current model, we can find the boundary conditions for this model,

$$\boldsymbol{n} \times (\boldsymbol{E}_1 - \boldsymbol{E}_2) = 0, \quad (4.4.7)$$

$$\boldsymbol{n} \times (\boldsymbol{H}_1 - \boldsymbol{H}_2) = 0, \quad (4.4.8)$$

$$\boldsymbol{n} \cdot (\boldsymbol{D}_1 - \boldsymbol{D}_2) = 0, \quad (4.4.9)$$

$$\boldsymbol{n} \cdot (\boldsymbol{B}_1 - \boldsymbol{B}_2) = 0, \quad (4.4.10)$$

where the definitions of subscripts attached to the field vectors and vector $\boldsymbol{n}$ are the same as for the convection current model.

4.5 Energy Conservation Relation for Small-Signal Fields

According to the polarization current model, a relativistic electron beam is macroscopically regarded as a bianisotropic medium with frequency and wave number dispersion. In the following discussion, on the basis of the polarization current model, we formulate the law of energy conservation for small-signal fields propagated along a relativistic electron beam, and define the energy density and the power flow density associated with small-signal fields. As clarified in Sect. 4.3, a relativistic electron beam is dispersive in terms of frequency, possessing material constants which are functions of frequency. Hence, as stated in Sect. 1.7, in order to formulate properly the law of energy conservation in a relativistic electron beam, we must treat quasi-sinusoidal fields with amplitudes slowly varying in time, instead of fields sinusoidally varying in time at a single frequency. At the same time, a relativistic electron beam is dispersive in terms of wave number as well, and its material constants become functions of wave number in the direction of beam propagation, namely, in the z direction. Hence, for a proper formulation of energy conservation relation in a relativistic electron beam, we must also consider quasi-sinusoidal fields with amplitudes slowly varying in the z direction.

Thus let us consider electromagnetic fields with amplitudes slowly varying in terms of t and z, whose electric field vector $\boldsymbol{E}(\boldsymbol{r}, t)$ is assumed to take the form

$$\boldsymbol{E}(\boldsymbol{r},t) = \mathrm{Re}\left[\tilde{\boldsymbol{E}}(\boldsymbol{r},t)e^{j(\omega t - k_z z)}\right], \tag{4.5.1}$$

where $\tilde{\boldsymbol{E}}(\boldsymbol{r},t)$ is a function of space and time with an amplitude slowly varying in t and z, which satisfies the conditions

$$\left|\frac{\partial \tilde{\boldsymbol{E}}}{\partial t}\right| << \omega\left|\tilde{\boldsymbol{E}}\right|, \qquad \left|\frac{\partial \tilde{\boldsymbol{E}}}{\partial z}\right| << k_z\left|\tilde{\boldsymbol{E}}\right|. \tag{4.5.2}$$

Other field vectors can also be represented in the same form as (4.5.1). Field vectors represented as (4.5.1) can be regarded as slightly modulated in terms of frequency and wave number in the z direction.

The complex amplitude of the electric field vector, $\tilde{\boldsymbol{E}}(\boldsymbol{r},t)$ defined by (4.5.1) can be expressed in terms of the Fourier integral as

$$\tilde{\boldsymbol{E}}(\boldsymbol{r},t) = \frac{1}{(2\pi)^2}\iint \overline{\boldsymbol{E}}(\omega',k_z')e^{j\left[(\omega'-\omega)t-(k_z'-k_z)z\right]}d\omega' dk_z', \tag{4.5.3}$$

where $\overline{\boldsymbol{E}}(\omega',k_z')$ is the Fourier component for electric field vector, which does not vanish only in the vicinity of $\omega' = \omega$ and $k_z' = k_z$. It should be noted that the Fourier component $\overline{\boldsymbol{E}}(\omega',k_z')$ is implicitly assumed to be a function of transverse spatial coordinates.

Noting that the constitutive relations (4.3.28) and (4.3.29) hold for each Fourier component of the field vectors, we find that the complex amplitude $\tilde{\boldsymbol{D}}(\boldsymbol{r},t)$ of the field vector $\boldsymbol{D}(\boldsymbol{r}, \mathrm{t})$ can be represented as

$$\tilde{\boldsymbol{D}}(\boldsymbol{r},t) = \frac{1}{(2\pi)^2}\iint \hat{\kappa}\cdot\overline{\boldsymbol{E}}(\omega',k_z')e^{j\left[(\omega'-\omega)t-(k_z'-k_z)z\right]}d\omega' dk_z'$$

$$+\frac{1}{(2\pi)^2}\iint \hat{\tau}\cdot\overline{\boldsymbol{H}}(\omega',k_z')e^{j\left[(\omega'-\omega)t-(k_z'-k_z)z\right]}d\omega' dk_z', \tag{4.5.4}$$

where $\overline{\boldsymbol{H}}(\omega',k_z')$ denotes the Fourier component for the complex amplitude of the field vector $\boldsymbol{H}(\boldsymbol{r}, t)$.

In (4.5.4), the Fourier components $\overline{\boldsymbol{E}}(\omega',k_z')$ and $\overline{\boldsymbol{H}}(\omega',k_z')$ for the complex amplitudes of the field vectors $\boldsymbol{E}(\boldsymbol{r},t)$ and $\boldsymbol{H}(\boldsymbol{r}, t)$ do not vanish only in the vicinity of $\omega' = \omega$ and $k_z' = k_z$. In view of this condition, Eq. (4.5.4) can be rewritten as

$$\tilde{\boldsymbol{D}}(r,t) = \hat{\kappa}\cdot\tilde{\boldsymbol{E}} - j\frac{\partial\hat{\kappa}}{\partial\omega}\cdot\frac{\partial\tilde{\boldsymbol{E}}}{\partial t} + j\frac{\partial\hat{\kappa}}{\partial k_z}\cdot\frac{\partial\tilde{\boldsymbol{E}}}{\partial z}$$

$$+\hat{\tau}\cdot\tilde{\boldsymbol{H}} - j\frac{\partial\hat{\tau}}{\partial\omega}\cdot\frac{\partial\tilde{\boldsymbol{H}}}{\partial t} + j\frac{\partial\hat{\tau}}{\partial k_z}\cdot\frac{\partial\tilde{\boldsymbol{H}}}{\partial z}, \tag{4.5.5}$$

where $\hat{\kappa}, \hat{\tau}$, and their derivatives with respect to ω and k_z are evaluated at $\omega' = \omega$

and $k_z' = k_z$. In a similar fashion, with the aid of (4.3.29), the complex amplitude of the field vector $\boldsymbol{B}(\boldsymbol{r}, t)$ can be represented in the same form as (4.5.5). Specifically, we can get the representation for $\tilde{\boldsymbol{B}}(\boldsymbol{r},t)$ by replacing $\hat{\kappa}$ and $\hat{\tau}$ on the right-hand side of (4.5.5) by $\hat{\tau}$ and $\hat{\mu}$, respectively.

From the basic field equations for the polarization current model, (4.2.20) and (4.2.21), we get the following relation:

$$\nabla\cdot\left[\boldsymbol{E}(\boldsymbol{r},t)\times\boldsymbol{H}(\boldsymbol{r},t)\right] + \left[\boldsymbol{E}(\boldsymbol{r},t)\cdot\frac{\partial\boldsymbol{D}(\boldsymbol{r},t)}{\partial t} + \boldsymbol{H}(\boldsymbol{r},t)\cdot\frac{\partial\boldsymbol{B}(\boldsymbol{r},t)}{\partial t}\right] + \boldsymbol{E}(\boldsymbol{r},t)\cdot\boldsymbol{J}_s(\boldsymbol{r},t) = 0. \tag{4.5.6}$$

where an externally applied source current $\boldsymbol{J}_s$ is included for generality. Substitute in (4.5.6) the relations (4.5.1) and (4.5.5), and the corresponding relations for the magnetic field vectors, and average temporally (4.5.6) over one period of time, $2\pi/\omega$ for quasi-sinusoidally varying fields. Then, we obtain the law of energy conservation for small-signal fields expressed in terms of the time-average energy density and power flow density as follows:

$$\nabla\cdot\langle\boldsymbol{S}\rangle + \frac{\partial\langle w\rangle}{\partial t} + \langle q\rangle = 0, \tag{4.5.7}$$

where

$$\langle\boldsymbol{S}\rangle = \langle\boldsymbol{S}_0\rangle + \langle\boldsymbol{S}_1\rangle, \tag{4.5.8}$$

$$\langle\boldsymbol{S}_0\rangle = \frac{1}{2}\mathrm{Re}\left[\tilde{\boldsymbol{E}}\times\tilde{\boldsymbol{H}}^*\right], \tag{4.5.9}$$

$$\langle\boldsymbol{S}_1\rangle = -\boldsymbol{i}_z\frac{1}{4}\left[\tilde{\boldsymbol{E}}^*\cdot\frac{\partial\omega\hat{\kappa}}{\partial k_z}\cdot\tilde{\boldsymbol{E}} + \tilde{\boldsymbol{H}}^*\cdot\frac{\partial\omega\hat{\mu}}{\partial k_z}\cdot\tilde{\boldsymbol{H}} + 2\,\mathrm{Re}\left(\tilde{\boldsymbol{E}}^*\cdot\frac{\partial\omega\hat{\tau}}{\partial k_z}\cdot\tilde{\boldsymbol{H}}\right)\right], \tag{4.5.10}$$

$$\langle w\rangle = \frac{1}{4}\left[\tilde{\boldsymbol{E}}^*\cdot\frac{\partial\omega\hat{\kappa}}{\partial\omega}\cdot\tilde{\boldsymbol{E}} + \tilde{\boldsymbol{H}}^*\cdot\frac{\partial\omega\hat{\mu}}{\partial\omega}\cdot\tilde{\boldsymbol{H}} + 2\,\mathrm{Re}\left(\tilde{\boldsymbol{E}}^*\cdot\frac{\partial\omega\hat{\tau}}{\partial\omega}\cdot\tilde{\boldsymbol{H}}\right)\right], \tag{4.5.11}$$

$$\langle q\rangle = \frac{1}{2}\mathrm{Re}\left(\tilde{\boldsymbol{E}}\cdot\tilde{\boldsymbol{J}}_s^*\right). \tag{4.5.12}$$

In the law of energy conservation for small-signal fields, (4.5.7), $\langle w\rangle$ and $\langle\boldsymbol{S}\rangle$ represent the time-average values of energy density and power flow density associated with small-signal fields propagated in a relativistic electron beam. In addition, $\langle q\rangle$ denotes the average power supplied from an external source per unit volume. The power flow density $\langle\boldsymbol{S}\rangle$ is composed of two parts, $\langle\boldsymbol{S}_0\rangle$ and $\langle\boldsymbol{S}_1\rangle$. The former $\langle\boldsymbol{S}_0\rangle$ is due to the Poynting vector, denoting the power flow density carried by electromagnetic fields themselves, while the latter $\langle\boldsymbol{S}_1\rangle$ expresses the power flow density carried by the drift of the electron beam, which is inherent to it.

The representation of $\langle \boldsymbol{S}_1 \rangle$ given by (4.5.10) suggests that the wave number dispersion or spatial dispersion in the electron beam as a macroscopic medium derives from its drift. To prove that $\langle \boldsymbol{S}_1 \rangle$ corresponds to the power flow due to the drift of the electron beam, we separate the energy density $\langle w \rangle$ into two terms, namely, one term independent of the dispersion of the electron beam and the additional term dependent on the dispersion as

$$\langle w \rangle = \langle w_0 \rangle + \langle w_1 \rangle, \tag{4.5.13}$$

where

$$\begin{aligned}\langle w_0 \rangle &= \frac{1}{4}\left[\tilde{\boldsymbol{E}}^* \cdot \hat{\kappa} \cdot \tilde{\boldsymbol{E}} + \tilde{\boldsymbol{H}}^* \cdot \hat{\mu} \cdot \tilde{\boldsymbol{H}} + 2\,\mathrm{Re}\left(\tilde{\boldsymbol{E}}^* \cdot \hat{\tau} \cdot \tilde{\boldsymbol{H}}\right)\right] \\ &= \frac{1}{4}\mathrm{Re}\left(\tilde{\boldsymbol{E}}^* \cdot \boldsymbol{D} + \tilde{\boldsymbol{H}}^* \cdot \boldsymbol{B}\right),\end{aligned} \tag{4.5.14}$$

$$\langle w_1 \rangle = \frac{1}{4}\omega\left[\tilde{\boldsymbol{E}}^* \cdot \frac{\partial \hat{\kappa}}{\partial \omega} \cdot \tilde{\boldsymbol{E}} + \tilde{\boldsymbol{H}}^* \cdot \frac{\partial \hat{\mu}}{\partial \omega} \cdot \tilde{\boldsymbol{H}} + 2\,\mathrm{Re}\left(\tilde{\boldsymbol{E}}^* \cdot \frac{\partial \hat{\tau}}{\partial \omega} \cdot \tilde{\boldsymbol{H}}\right)\right]. \tag{4.5.15}$$

It should be remembered here that the material constants of a relativistic electron beam, namely, the elements of the tensors $\hat{\kappa}$, $\hat{\mu}$, and $\hat{\tau}$ depend on ω and k_z only through $\omega' = \gamma(\omega - v_0 k_z)$. Then, we readily find the following relations:

$$\frac{\partial \hat{\kappa}}{\partial k_z} = -v_0 \frac{\partial \hat{\kappa}}{\partial \omega}, \quad \frac{\partial \hat{\mu}}{\partial k_z} = -v_0 \frac{\partial \hat{\mu}}{\partial \omega}, \quad \frac{\partial \hat{\tau}}{\partial k_z} = -v_0 \frac{\partial \hat{\tau}}{\partial \omega}. \tag{4.5.16}$$

Comparing $\langle \boldsymbol{S}_1 \rangle$ in (4.5.10) with $\langle w_1 \rangle$ in (4.5.15), with the aid of (4.5.16), we get the relation

$$\langle \boldsymbol{S}_1 \rangle = \langle w_1 \rangle \boldsymbol{v}_0. \tag{4.5.17}$$

The above equation means that the power flow density $\langle \boldsymbol{S}_1 \rangle$ corresponds to the additional energy density due to the dispersion of the electron beam, $\langle w_1 \rangle$ carried by its drift. In the theory of electron beams, we should note an important role played by the power flow carried by their drift, or the power flow associated with their spatial dispersion.

The law of energy conservation for small-signal fields in a relativistic electron beam given by (4.5.7) represents a local law which holds at a particular point in it. However, it is the law of energy conservation for unit length of the electron beam that the theory of electron beams actually requires. In order to find the law of energy conservation for unit length of the electron beam, we first integrate (4.5.7) in the region enclosed by two planes normal to the direction of beam flow, which is separated by Δz, and a metallic wall surrounding the beam. Then, we make Δz tend to zero. Thus we get the following expression for the law of energy conservation for unit length of the electron beam:

$$\frac{\partial\langle P\rangle}{\partial z}+\frac{\partial\langle W\rangle}{\partial t}+\langle Q\rangle=0, \tag{4.5.18}$$

with

$$\langle P\rangle=\int\langle \boldsymbol{S}\rangle\cdot\boldsymbol{i}_z ds=\langle P_0\rangle+\langle P_1\rangle, \tag{4.5.19}$$

$$\langle P_0\rangle=\int\langle \boldsymbol{S}_0\rangle\cdot\boldsymbol{i}_z ds,\quad \langle P_1\rangle=\int\langle \boldsymbol{S}_1\rangle\cdot\boldsymbol{i}_z ds, \tag{4.5.20}$$

$$\langle W\rangle=\int\langle w\rangle ds=\langle W_0\rangle+\langle W_1\rangle, \tag{4.5.21}$$

$$\langle W_0\rangle=\int\langle w_0\rangle ds,\quad \langle W_1\rangle=\int\langle w_1\rangle ds, \tag{4.5.22}$$

$$\langle Q\rangle=\int\langle q\rangle ds, \tag{4.5.23}$$

where the surface integration is carried out on the plane normal to the electron beam, and $\langle P\rangle$ and $\langle W\rangle$ denote, respectively, the power flow along the electron beam and the density of field energy per unit length of it. In addition, $\langle Q\rangle$ is the power supplied from an external source per unit length of the electron beam. The average quantities $\langle P_0\rangle$ and $\langle W_0\rangle$ express the power flow and energy density independent of the beam dispersion, while $\langle P_1\rangle$ and $\langle W_1\rangle$ correspond to the additional power flow and energy density due to the beam dispersion. It should be noted that $\langle P\rangle$ and $\langle W\rangle$ are slowly varying functions of z and t.

The law of energy conservation for unit length of the electron beam (4.5.18) was derived on the assumption that the electron beam was surrounded by a metallic wall. However, it holds for an electron beam isolated in unbounded space. For this case, the surface integration in (4.5.19) ~ (4.5.23) must be carried out over an infinite plane perpendicular to the electron beam.

4.6 Group Velocity and Energy Transport Velocity

The small-signal fields propagated along the electron beam are generally expressed as (4.5.1). Then, it should be noted that the relation between ω and k_z, namely, the dispersion relation is determined from the basic field equations to describe the small-signal fields propagated along the electron beam and the boundary conditions. First, let us consider an electromagnetic wave with single frequency ω and single wave number k_z. The plane of constant phase for this wave is propagated with the velocity $v_p = \omega/k_z$ in the z direction, which is referred to as the phase velocity. In electron beams, the phase velocity is generally different from those velocities with which signal and energy are carried.

In the following discussion, we consider those velocities with which signal (information) and energy are propagated. The aforementioned electromagnetic wave with single frequency and single wave number lasts for ever continuously in the temporal and spatial domains $-\infty < t < \infty$ and $-\infty < z < \infty$. Such a wave can not propagate any kind of signal or information. In other words, in order to transmit signal or information, one must modulate a carrier wave with single frequency ω and single wave number k_z, in accordance with signal or information. A modulated carrier wave no more has single frequency and single wave number. Instead, it comes to have frequency and wave number spectra spread around the frequency and wave number of the carrier wave. According to the Fourier analysis, such a modulated wave is regarded as composed of a large number of waves with single frequencies and single wave numbers. A group of large numbers of waves with single frequencies and single wave numbers, which constitute narrow spectra around the central frequency and wave number, is referred to as a wave packet. In other words, signal or information is carried by a wave packet, and that velocity with which it is propagated is called the group velocity. If a wave packet is to be uniquely defined and the signal carried by it is to be transmitted without distortion, frequencies and wave numbers of Fourier components constituting a wave packet must be distributed over narrow spectral ranges around the frequency and wave number of the carrier wave.

Let us consider a wave packet around a particular point (ω, k_z) on a dispersion curve. Then, $\tilde{\boldsymbol{E}}(\boldsymbol{r},t)$ in (4.5.1) is expressed as

$$\tilde{\boldsymbol{E}}(\boldsymbol{r},t) = \frac{1}{2\pi}\int \overline{\boldsymbol{E}}(k_z')e^{j\left[\{\omega(k_z')-\omega(k_z)\}t-(k_z'-k_z)z\right]}dk_z', \tag{4.6.1}$$

where $\overline{\boldsymbol{E}}(k_z')$ has a sharp peak at $k_z' = k_z$, not vanishing only in a small region around it. For this reason, after being expanded in a Taylor series around $k_z' = k_z$, $\omega(k_z')$ can be approximated by the following two terms:

$$\omega(k_z') = \omega(k_z) + \frac{\partial\omega}{\partial k_z}\left(k_z' - k_z\right), \tag{4.6.2}$$

where $\partial\omega / \partial k_z$ is evaluated at $k_z' = k_z$.

Using the relation (4.6.2) in (4.6.1), we get

$$\tilde{\boldsymbol{E}}(\boldsymbol{r},t) = \boldsymbol{A}\left(\frac{\partial\omega}{\partial k_z}t - z\right), \tag{4.6.3}$$

where

$$\boldsymbol{A}\left(\frac{\partial\omega}{\partial k_z}t - z\right) = \frac{1}{2\pi}\int \overline{\boldsymbol{E}}(k_z')e^{j\left[(k_z'-k_z)\left(\frac{\partial\omega}{\partial k_z}t-z\right)\right]}dk_z'. \tag{4.6.4}$$

The function $\boldsymbol{A}$ is a slowly varying function of t and z. Now, let us consider a plane perpendicular to the z-axis for which

$$\frac{\partial\omega}{\partial k_z}t - z = \text{constant},$$

and then we find that this plane translates with the velocity

$$v_g = \frac{\partial\omega}{\partial k_z}, \tag{4.6.5}$$

in the z direction. The velocity defined by (4.6.5) is referred to as the group velocity. If the group velocity is to be uniquely defined, the bandwidth covered by a wave packet is so narrow that the inclination of the tangent to a dispersion curve can be regarded as constant within it. As is evident from the above explanation, the group velocity is equal to the velocity with which the envelope of a modulated wave is propagated, namely, signal or information is transmitted.

Next, let us investigate the relation between the group velocity and the energy transport velocity. For this purpose, we assume that the angular frequency of a wave has an imaginary part $j\omega_i$, which is vanishingly small compared with the real part ω. Then, from (4.6.2), we find that the wave number also has an imaginary part $j(\omega_i/v_g)$ much less than the real part k_z. Hence, for this case, $\boldsymbol{A}$ takes, from (4.6.4), the form

$$\boldsymbol{A}(v_g t - z) = \boldsymbol{E}_0 \exp\left[-\omega_i\left(t - \frac{z}{v_g}\right)\right], \tag{4.6.6}$$

where $\boldsymbol{E}_0$ denotes a spatial vector independent of t and z.

From (4.5.8) ~ (4.511) and (4.5.19) ~ (4.5.22), the time-average values for the power flow propagated along the electron beam, $\langle P\rangle$ and the energy density for unit length of it, $\langle W\rangle$ are found to be proportional to the square of the absolute value of $\boldsymbol{A}$. Hence, from (4.5.18) with $\langle Q\rangle = 0$, we find the relation

$$v_g = \frac{\partial\omega}{\partial k_z} = \frac{\langle P\rangle}{\langle W\rangle}, \tag{4.6.7}$$

where the right-hand side $\langle P\rangle/\langle W\rangle$ represents the velocity with which energy is transported along the electron beam. From the above discussion, we can conclude that the energy transport velocity is equal to the group velocity for the case where the dissipation in the electron beam can be neglected.

4.7 Transformation of Energy Density and Power Flow

In Sect. 4.5, we formulated the law of energy conservation for small-signal fields propagated along a relativistic electron beam, and defined the power flow carried by it and the energy density for unit length of it. In this section, we derive the transformation formulas for the power flow and the energy density between the laboratory frame and the rest frame of the electron beam. In the following discussion, in order to find the transformation formulas for the power flow and the energy density between the two inertial systems, we use a method based upon the concept of quasi-particles [4.15, 4.21]. This method, although simple, has general applicability.

From a quantum-mechanical point of view, a wave packet with the center frequency ω and the center wave number k_z in the z direction in a macroscopic medium can be regarded as an aggregate of quasi-particles with energy $\hbar\omega$ and momentum $\hbar k_z$. Here, $\hbar$ is equal to the Planck constant h divided by 2π. Hence, from a standpoint of quantum mechanics, the energy density $\langle W \rangle$ and the power flow $\langle P \rangle$ associated with small-signal fields propagated along the electron beam can be represented as

$$\langle W \rangle = \hbar\omega \langle N_\omega \rangle, \tag{4.7.1}$$

$$\langle P \rangle = \hbar\omega \langle N_\omega \rangle v_g, \tag{4.7.2}$$

where $\langle N_\omega \rangle$ denotes the average number density of quasi-particles per unit length of the electron beam, and v_g is the group velocity defined in the preceding section.

From (4.7.1) and (4.7.2), let us define new physical quantities as

$$\frac{\langle W \rangle}{\omega} = \hbar \langle N_\omega \rangle, \tag{4.7.3}$$

$$\frac{\langle P \rangle}{\omega} = \hbar \langle N_\omega \rangle v_g. \tag{4.7.4}$$

In the above equations, noting that $|v_g| < c$ and $\hbar$ is a constant, we can compare $\hbar$ to the charge of a charged particle, $\langle N_\omega \rangle$ to the number density of charged particles per unit length, and v_g to their velocity, respectively. Hence, the physical quantities $\langle W \rangle / \omega$ and $\langle P \rangle / \omega$ divided by the cross section of the electron beam can be compared to the charge density and the current density, respectively. It should be noted that the cross section of the electron beam is kept invariant under the Lorentz transformation. Thus, from the above comparison, we find that the transformation formulas for $\langle W \rangle / \omega$ and $\langle P \rangle / \omega$ can be obtained from those for charge and current densities as

$$\frac{\langle W' \rangle}{\omega'} = \gamma \left[\frac{\langle W \rangle}{\omega} - \frac{v_0}{c^2} \frac{\langle P \rangle}{\omega} \right], \tag{4.7.5}$$

$$\frac{\langle P'\rangle}{\omega'} = \gamma\left[\frac{\langle P\rangle}{\omega} - v_0\frac{\langle W\rangle}{\omega}\right], \tag{4.7.6}$$

where $\langle W\rangle$ and $\langle P\rangle$ are the energy density and the power flow in the laboratory frame, and $\langle W'\rangle$ and $\langle P'\rangle$ the corresponding quantities in the rest frame of the electron beam. In addition, the relation between ω' and ω is given by

$$\omega' = \gamma\left(\omega - v_0 k_z\right), \tag{4.7.7}$$

which is obtained from (2.5.5).

Inserting the relation (4.6.7) in (4.7.5) and (4.7.6), we get

$$\frac{\langle W'\rangle}{\omega'} = \gamma\left(1 - \frac{v_0 v_g}{c^2}\right)\frac{\langle W\rangle}{\omega}, \tag{4.7.8}$$

$$\frac{\langle P'\rangle}{\omega'} = \gamma\left(1 - \frac{v_0}{v_g}\right)\frac{\langle P\rangle}{\omega}. \tag{4.7.9}$$

Let us discuss the properties of the energy density and the power flow obtained from the transformation formulas for them. First, with the aid of (4.7.7), Eq. (4.7.8) can be rewritten as

$$\langle W'\rangle = \gamma^2\left(1 - \frac{v_0 v_g}{c^2}\right)\left(\omega - v_0 k_z\right)\frac{\langle W\rangle}{\omega}. \tag{4.7.10}$$

Here, we express the energy density in the rest frame of the electron beam, $\langle W'\rangle$ in terms of field vectors and the constitutive relations in that frame. The constitutive relations in the rest frame of the electron beam can be obtained by setting $v_0 = 0$ in (4.3.28) and (4.3.29) as

$$\boldsymbol{D}' = \varepsilon_0\hat{\varepsilon}\boldsymbol{E}', \quad \boldsymbol{B}' = \mu_0\boldsymbol{H}', \tag{4.7.11}$$

where $\hat{\varepsilon}$ is the permittivity tensor in the rest frame of the electron beam, which is defined in (4.3.20), and the primed field vectors refer to those in that frame.

In the rest frame of the electron beam, it should be noted that we have $\hat{\kappa} = \varepsilon_0\hat{\varepsilon}, \hat{\tau} = 0$, and $\hat{\mu} = \mu_0\hat{u}$ from (4.3.31) ~ (4.3.33). Then, the energy density is expressed from (4.5.11) and (4.5.21) as

$$\langle W'\rangle = \frac{1}{4}\varepsilon_0\int \tilde{\boldsymbol{E}}'^* \cdot \frac{\partial\omega'\hat{\varepsilon}}{\partial\omega'}\cdot\tilde{\boldsymbol{E}}'ds + \frac{1}{4}\mu_0\int\tilde{\boldsymbol{H}}'^*\cdot\tilde{\boldsymbol{H}}'ds. \tag{4.7.12}$$

Substituting (4.3.20) for $\hat{\varepsilon}$ in (4.7.12), we find that $\langle W'\rangle > 0$ always holds. Namely, the energy density always takes a positive value in the rest frame of the electron beam. On the other hand, in the laboratory frame in which the electron beam is drifting with the velocity v_0, the energy density $\langle W\rangle$ can take a negative value under a particular condition. The condition for $\langle W\rangle$ to be negative can be found from

(4.7.10). First of all, the left-hand side of (4.7.10) is always positive, as shown in the foregoing discussion. On the other hand, the factor $(1- v_0 v_g/c^2)$ on the right-hand side is always positive, because the drift velocity of the electron beam and the group velocity of small-signal fields propagated along it are always less than the vacuum velocity of light, c. In addition, we naturally have $\omega > 0$. Then, if we have the following condition satisfied:

$$\omega - v_0 k_z < 0, \tag{4.7.13}$$

then the energy density can take a negative value in the laboratory frame. Noting that ω/k_z is equal to the phase velocity v_p, we can rewrite the condition (4.7.13) as

$$v_p < v_0. \tag{4.7.14}$$

Hence the energy density becomes negative for the case where the phase velocity is less than the drift velocity of the electron beam. On the other hand, for the case

$$\omega - v_0 k_z > 0, \tag{4.7.15}$$

or

$$v_p > v_0, \tag{4.7.16}$$

the energy density becomes positive.

Next, Eq. (4.7.9) can be reduced to

$$\langle P' \rangle = \gamma^2 \left(1 - \frac{v_0}{v_g}\right)\left(1 - \frac{v_0}{v_p}\right)\langle P \rangle. \tag{4.7.17}$$

Let us assume $v_p > 0$ and $v_g > 0$. Then, $\langle P' \rangle$ and $\langle P \rangle$ take different signs for the following two cases:

$$(1) \quad v_p > v_g, \quad v_g < v_0 < v_p, \tag{4.7.18}$$

$$(2) \quad v_p < v_g, \quad v_p < v_0 < v_g, \tag{4.7.19}$$

while they take the same signs when v_0 falls outside the above ranges.

In the above discussion, we referred to the negative energy density associated with small-signal fields propagated along the electron beam. Here, let us clarify the physical meaning of positive and negative energies carried by small-signal fields in the electron beam. First of all, in the absence of small-signal fields in the electron beam, it carries constant kinetic energy of density W_b, corresponding to its uniform drift or translational motion. Now, let small-signal fields be excited in it. Then, if the total energy density of the whole system composed of the electron beam and small-signal fields becomes greater than W_b, small-signal fields carry positive energy. On the other hand, if the whole system carries energy less than W_b, the energy of small-signal fields is negative. This is the physical meaning of positive and negative energies transported by small-signal fields in the electron beam. The waves in the electron beam transporting negative energy can be coupled with various types of positive-energy waves to generate or amplify high-power electromagnetic waves ranging from centimeter to submillimeter waves.

4.8 Momentum Conservation Relation for Small-Signal Fields

Let us first derive a vector identity for local and instantaneous values of electromagnetic fields from the Maxwell equations in the polarization current model, which constitutes a starting point of the discussion on momentum conservation relation for small-signal fields propagated along a relativistic electron beam. For this purpose, we consider the Lorentz force acting on externally applied charge and current distributions in a material medium. From the Maxwell equations in the polarization current model, (4.2.20) ~ (4.2.23) supplemented with charge density ρ_s and current density $\boldsymbol{J}_s$, we get the following relation:

$$\rho_s \boldsymbol{E} + \boldsymbol{J}_s \times \boldsymbol{B} = (\nabla \cdot \boldsymbol{D})\boldsymbol{E} + \left(\nabla \times \boldsymbol{H} - \frac{\partial \boldsymbol{D}}{\partial t}\right) \times \boldsymbol{B}$$

$$+(\nabla \cdot \boldsymbol{B})\boldsymbol{H} + \left(\nabla \times \boldsymbol{E} + \frac{\partial \boldsymbol{B}}{\partial t}\right) \times \boldsymbol{D}, \tag{4.8.1}$$

where the third and fourth terms on the right-hand side, which are identically zero, are added for the subsequent manipulations. Equation (4.8.1) can be rearranged as

$$\rho_s \boldsymbol{E} + \boldsymbol{J}_s \times \boldsymbol{B} + \frac{\partial}{\partial t}(\boldsymbol{D} \times \boldsymbol{B}) = (\nabla \cdot \boldsymbol{D})\boldsymbol{E} - \boldsymbol{D} \times (\nabla \times \boldsymbol{E})$$

$$+(\nabla \cdot \boldsymbol{B})\boldsymbol{H} - \boldsymbol{B} \times (\nabla \times \boldsymbol{H}). \tag{4.8.2}$$

In order to rewrite the first and second terms on the right-hand side of (4.8.2), we note the following vector identities:

$$\nabla(\boldsymbol{D} \cdot \boldsymbol{E}) = \boldsymbol{E} \times (\nabla \times \boldsymbol{D}) + \boldsymbol{D} \times (\nabla \times \boldsymbol{E}) + (\boldsymbol{E} \cdot \nabla)\boldsymbol{D} + (\boldsymbol{D} \cdot \nabla)\boldsymbol{E}, \tag{4.8.3}$$

$$\nabla \cdot (\boldsymbol{D}\boldsymbol{E}) = (\nabla \cdot \boldsymbol{D})\boldsymbol{E} + (\boldsymbol{D} \cdot \nabla)\boldsymbol{E}. \tag{4.8.4}$$

With the aid of these vector identities, the first and second terms on the right-hand side of (4.8.2) can be reduced to

$$(\nabla \cdot \boldsymbol{D})\boldsymbol{E} - \boldsymbol{D} \times (\nabla \times \boldsymbol{E}) = \nabla \cdot \left[\boldsymbol{D}\boldsymbol{E} - \left(\frac{1}{2}\boldsymbol{D} \cdot \boldsymbol{E}\right)\ddot{u}\right]$$

$$-\frac{1}{2}\left[(\nabla \boldsymbol{E}) \cdot \boldsymbol{D} - (\nabla \boldsymbol{D}) \cdot \boldsymbol{E}\right], \tag{4.8.5}$$

$\ddot{u}$ being a unit dyadic. The third and fourth terms on the right-hand side of (4.8.2) can be rewritten in a similar fashion. Then, Eq. (4.8.2) can be finally reduced to

$$\nabla\cdot\left[\frac{1}{2}(\boldsymbol{D}\cdot\boldsymbol{E}+\boldsymbol{B}\cdot\boldsymbol{H})\ddot{u}-(\boldsymbol{DE}+\boldsymbol{BH})\right]+\frac{\partial}{\partial t}(\boldsymbol{D}\times\boldsymbol{B})$$
$$+\frac{1}{2}\left[(\nabla\boldsymbol{E})\cdot\boldsymbol{D}-(\nabla\boldsymbol{D})\cdot\boldsymbol{E}\right]+\frac{1}{2}\left[(\nabla\boldsymbol{H})\cdot\boldsymbol{B}-(\nabla\boldsymbol{B})\cdot\boldsymbol{H}\right]$$
$$=-(\rho_s\boldsymbol{E}+\boldsymbol{J}_s\times\boldsymbol{B}), \tag{4.8.6}$$

The second line of (4.8.6) vanishes for a special case of homogeneous dispersion-free medium.

Taking the z component of (4.8.6) and integrating it over a plane perpendicular to the direction of beam flow, we get

$$\frac{\partial}{\partial z}\frac{1}{2}\int(\boldsymbol{D}_t\cdot\boldsymbol{E}_t+\boldsymbol{B}_t\cdot\boldsymbol{H}_t-\boldsymbol{D}_z\cdot\boldsymbol{E}_z+\boldsymbol{B}_z\cdot\boldsymbol{H}_z)ds$$
$$+\frac{\partial}{\partial t}\int(\boldsymbol{D}\times\boldsymbol{B})\cdot\boldsymbol{i}_z ds+\frac{1}{2}\int\left(\frac{\partial\boldsymbol{E}}{\partial z}\cdot\boldsymbol{D}-\frac{\partial\boldsymbol{D}}{\partial z}\cdot\boldsymbol{E}\right)ds$$
$$+\frac{1}{2}\int\left(\frac{\partial\boldsymbol{H}}{\partial z}\cdot\boldsymbol{B}-\frac{\partial\boldsymbol{B}}{\partial z}\cdot\boldsymbol{H}\right)ds=-\int(\rho_s\boldsymbol{E}+\boldsymbol{J}_s\times\boldsymbol{B})\cdot\boldsymbol{i}_z ds, \tag{4.8.7}$$

where subscripts t and z denote the transverse and longitudinal field components.

Considering a slightly modulated wave packet as used in Sect. 4.5, let us average (4.8.7) over one period of the rapidly varying part of the wave packet. Then, we obtain the following momentum conservation law in terms of time-average quantities:

$$\frac{\partial\langle T\rangle}{\partial z}+\frac{\partial\langle G\rangle}{\partial t}+\langle F\rangle=0, \tag{4.8.8}$$

with

$$\langle T\rangle=\langle T_0\rangle+\langle T_1\rangle, \tag{4.8.9}$$

$$\langle T_0\rangle=\mathrm{Re}\left[\frac{1}{4}\int\left(\tilde{\boldsymbol{D}}_t^*\cdot\tilde{\boldsymbol{E}}_t+\tilde{\boldsymbol{B}}_t^*\cdot\tilde{\boldsymbol{H}}_t-\tilde{\boldsymbol{D}}_z^*\cdot\tilde{\boldsymbol{E}}_z-\tilde{\boldsymbol{B}}_z^*\cdot\tilde{\boldsymbol{H}}_z\right)ds\right], \tag{4.8.10}$$

$$\langle T_1\rangle=-\frac{1}{4}k_z\int\left[\tilde{\boldsymbol{E}}^*\cdot\frac{\partial\hat{\kappa}}{\partial k_z}\cdot\tilde{\boldsymbol{E}}+\tilde{\boldsymbol{H}}^*\cdot\frac{\partial\hat{\mu}}{\partial k_z}\cdot\tilde{\boldsymbol{H}}+2\,\mathrm{Re}\left(\tilde{\boldsymbol{E}}^*\cdot\frac{\partial\hat{\tau}}{\partial k_z}\cdot\tilde{\boldsymbol{H}}\right)\right]ds, \tag{4.8.11}$$

$$\langle G\rangle=\langle G_0\rangle+\langle G_1\rangle, \tag{4.8.12}$$

$$\langle G_0\rangle=\frac{1}{2}\mathrm{Re}\left[\int\left(\tilde{\boldsymbol{D}}^*\times\tilde{\boldsymbol{B}}\right)\cdot\boldsymbol{i}_z ds\right], \tag{4.8.13}$$

$$\langle G_1 \rangle = \frac{1}{4} k_z \int \left[\tilde{\boldsymbol{E}}^* \cdot \frac{\partial \hat{\kappa}}{\partial \omega} \cdot \tilde{\boldsymbol{E}} + \tilde{\boldsymbol{H}}^* \cdot \frac{\partial \hat{\mu}}{\partial \omega} \cdot \tilde{\boldsymbol{H}} + 2 \operatorname{Re}\left(\tilde{\boldsymbol{E}}^* \cdot \frac{\partial \hat{\tau}}{\partial \omega} \cdot \tilde{\boldsymbol{H}} \right) \right] ds, \tag{4.8.14}$$

$$\langle F \rangle = \frac{1}{2} \operatorname{Re}\left[\int \left(\tilde{\rho}_s^* \tilde{\boldsymbol{E}} + \tilde{\boldsymbol{J}}_s \times \tilde{\boldsymbol{B}} \right) \cdot \boldsymbol{i}_z ds \right], \tag{4.8.15}$$

where $\langle G \rangle$ and $\langle F \rangle$ are the time-average momentum density and force density per unit length of the electron beam, and $\langle T \rangle$ the time-average momentum flow parallel to the beam flow. The subscripts 0 and 1 attached to T and G denote their nondispersive and dispersive parts, respectively. The momentum flow $\langle T_0 \rangle$ is carried by small-signal fields themselves, while the momentum flow $\langle T_1 \rangle$ is carried by the drift of the electron beam.

In the same manner as in Sect. 4.5, we can prove the relation

$$\langle T_1 \rangle = \langle G_1 \rangle v_0. \tag{4.8.16}$$

Thus the dispersive part of the momentum flow corresponds to the dispersive part of the momentum density carried with the drift of the electron beam. In addition, from the momentum conservation relation (4.8.8), with $\langle F \rangle = 0$, the group velocity of the small-signal fields is found to have an alternative representation

$$v_g = \frac{\langle T \rangle}{\langle G \rangle}. \tag{4.8.17}$$

In Sect. 4.5 and this section, we have discussed separately the energy and momentum relations for the small-signal fields propagated along a relativistic electron beam. However, it should be noted that the energy or power flow is not independent of the momentum or its flow but they are related to each other by a simple relationship. Namely, the momentum density $\langle G \rangle$ is connected with the energy density $\langle W \rangle$ through

$$\frac{\langle G \rangle}{k_z} = \frac{\langle W \rangle}{\omega}. \tag{4.8.18}$$

To prove this relation, we first note that the nondispersive parts of $\langle G \rangle$ and $\langle W \rangle$ satisfy

$$\frac{\langle G_0 \rangle}{k_z} = \frac{\langle W_0 \rangle}{\omega}, \tag{4.8.19}$$

which can be obtained directly from the source-free Maxwell equations, with the aid of the complex representation as (4.5.1). On the other hand, from (4.5.15), (4.5.22), and (4.8.14), the dispersive parts of $\langle G \rangle$ and $\langle W \rangle$ are readily found to satisfy

$$\frac{\langle G_1 \rangle}{k_z} = \frac{\langle W_1 \rangle}{\omega}, \tag{4.8.20}$$

which is combined with (4.8.19) to directly lead to (4.8.18).

Similarly, the momentum flow $\langle T \rangle$ and the power flow $\langle P \rangle$ are related to each other through

$$\frac{\langle T \rangle}{k_z} = \frac{\langle P \rangle}{\omega}, \tag{4.8.21}$$

which also yields from the fact that both the dispersive and nondispersive parts of $\langle T \rangle$ and $\langle P \rangle$ satisfy the same type of relation as (4.8.21). The relation between the nondispersive parts of $\langle T \rangle$ and $\langle P \rangle$ can be derived directly from the source-free Maxwell equations in the same manner as (4.8.19) was obtained, while the relation between their dispersive parts can be easily found from (4.5.10), (4.5.20), and (4.8.11).

4.9 Transformation of Momentum Density and Momentum Flow

The transformation formulas for momentum density and momentum flow between the laboratory frame and the rest frame of the electron beam can be obtained from those for energy density and power flow, with aid of the relations (4.8.18) and (4.8.21). However, we discuss separately the former transformation formulas on the basis of the concept of quasi-particles, which has already been used in deriving the transformation laws for energy density and power flow in Sect. 4.7. From the viewpoint of quasi-particles, the momentum density $\langle G \rangle$ and the momentum flow $\langle T \rangle$ can be represented as

$$\langle G \rangle = \hbar k_z \langle N_\omega \rangle, \tag{4.9.1}$$

$$\langle T \rangle = \hbar k_z \langle N_\omega \rangle v_g, \tag{4.9.2}$$

from which we can define new physical quantities

$$\frac{\langle G \rangle}{k_z} = \hbar \langle N_\omega \rangle, \tag{4.9.3}$$

$$\frac{\langle T \rangle}{k_z} = \hbar \langle N_\omega \rangle v_g. \tag{4.9.4}$$

Comparing (4.9.3) and (4.9.4) with (4.7.3) and (4.7.4), we find that $\langle G \rangle / k_z$ and $\langle T \rangle / k_z$ are equal, respectively, to $\langle W \rangle / \omega$ and $\langle P \rangle / \omega$. Thus, from the viewpoint of quasi-particles, the relations (4.8.18) and (4.8.21) can be easily obtained, and referring to (4.7.5) and (4.7.6), we readily get for the transformation formulas for momentum density and momentum flow,

$$\frac{\langle G'\rangle}{k_z'} = \gamma\left(\frac{\langle G\rangle}{k_z} - \frac{v_0}{c^2}\frac{\langle T\rangle}{k_z}\right), \tag{4.9.5}$$

$$\frac{\langle T'\rangle}{k_z'} = \gamma\left(\frac{\langle T\rangle}{k_z} - v_0\frac{\langle G\rangle}{k_z}\right), \tag{4.9.6}$$

where

$$k_z' = \gamma\left(k_z - \frac{v_0}{c^2}\omega\right), \tag{4.9.7}$$

which is obtained from (2.5.5).

With the aid of (4.8.17), Eqs. (4.9.5) and (4.9.6) are reduced to

$$\frac{\langle G'\rangle}{k_z'} = \gamma\left(1 - \frac{v_0 v_g}{c^2}\right)\frac{\langle G\rangle}{k_z}, \tag{4.9.8}$$

$$\frac{\langle T'\rangle}{k_z'} = \gamma\left(1 - \frac{v_0}{v_g}\right)\frac{\langle T\rangle}{k_z}, \tag{4.9.9}$$

which can be further rewritten, with the aid of (4.9.7), as

$$\langle G'\rangle = \gamma^2\left(1 - \frac{v_0 v_g}{c^2}\right)\left(1 - \frac{v_0 v_p}{c^2}\right)\langle G\rangle, \tag{4.9.10}$$

$$\langle T'\rangle = \gamma^2\left(1 - \frac{v_0}{v_g}\right)\left(1 - \frac{v_0 v_p}{c^2}\right)\langle T\rangle, \tag{4.9.11}$$

where v_p denotes the phase velocity defined as ω/k_z.

For slow beam waves for which $v_p < c$, as is evident from (4.9.10) and (4.9.11), the momentum density has the same sign in both the laboratory frame and the rest frame of the electron beam, whereas the momentum flow changes its sign for $v_0 > v_g$ when transformed from one frame to the other.

4.10 Waves in Relativistic Electron Beams

In this section, we discuss wave propagation in an unbounded relativistic electron beam on the basis of the convection current model for the electron beam. In particular, we consider only waves propagated in the direction of beam flow, which is very useful for various applications. From basic equations for the electron beam described in Sect. 4.2, we find that waves propagated parallel to an unbounded beam can be separated into two types of waves: one is a transverse wave consisting of field components perpendicular to the direction of beam flow, and the other is a longitudinal wave composed of field components parallel to it. In a general unbounded beam immersed in a finite magnetostatic field, two kinds of transverse waves, namely, electromagnetic waves and electron cyclotron waves can be propagated. Both of transverse waves are composed of right-hand and left-hand circularly polarized waves. The right-hand or left-hand circularly polarized wave refers to the wave whose electric field vector rotates clockwise or counterclockwise when looking in the direction of propagation. On the other hand, the longitudinal wave propagated in an unbounded beam is called a space-charge wave or electron plasma wave, properties of which are not affected by the applied magnetostatic field. In the following discussion, we investigate in detail the properties of transverse and longitudinal waves.

4.10.1 Electromagnetic Waves and Electron Cyclotron Waves

Let the electron beam be drifting in the z direction, and an arbitrary component of small-signal fields propagated in the z direction be expressed as

$$F(z,t) = \mathrm{Re}\left[\tilde{F}e^{j(\omega t - k_z z)}\right]. \tag{4.10.1}$$

Then, from (4.2.6), (4.2.7), and (4.2.4), we get for transverse waves in an unbounded beam,

$$k_z\tilde{E}_y = -\omega\tilde{B}_x, \quad k_z\tilde{E}_x = \omega\tilde{B}_y, \tag{4.10.2}$$

$$jk_z\tilde{B}_y = j\frac{\omega}{c^2}\tilde{E}_x + \mu_0\tilde{J}_x, \quad -jk_z\tilde{B}_x = j\frac{\omega}{c^2}\tilde{E}_y + \mu_0\tilde{J}_y, \tag{4.10.3}$$

$$\tilde{J}_x = -N_0 e\tilde{v}_x, \quad \tilde{J}_y = -N_0 e\tilde{v}_y, \tag{4.10.4}$$

where $\tilde{v}_x$ and $\tilde{v}_y$ satisfy, from (4.2.10),

$$j(\omega - v_0 k_z)\tilde{v}_x = -\frac{e}{\gamma m_0}\left(\tilde{E}_x - v_0\tilde{B}_y + B_0\tilde{v}_y\right), \tag{4.10.5}$$

$$j(\omega - v_0 k_z)\tilde{v}_y = -\frac{e}{\gamma m_0}\left(\tilde{E}_y + v_0\tilde{B}_x - B_0\tilde{v}_x\right), \tag{4.10.6}$$

$$\gamma = \frac{1}{\sqrt{1-\beta_0^2}}, \quad \beta_0 = \frac{v_0}{c}. \tag{4.10.7}$$

In order to treat circularly polarized waves, if we define new field components as

$$\tilde{E}_{\pm} = \tilde{E}_x \pm j\tilde{E}_y, \quad \tilde{B}_{\pm} = \tilde{B}_x \pm j\tilde{B}_y, \tag{4.10.8}$$

$$\tilde{J}_{\pm} = \tilde{J}_x \pm j\tilde{J}_y, \quad \tilde{v}_{\pm} = \tilde{v}_x \pm j\tilde{v}_y, \tag{4.10.9}$$

then Eqs. (4.10.2) ~ (4.10.6) can be reduced to

$$k_z\tilde{E}_{\pm} = \mp j\omega\tilde{B}_{\pm}, \tag{4.10.10}$$

$$\pm k_z\tilde{B}_{\pm} = j\frac{\omega}{c^2}\tilde{E}_{\pm} + \mu_0\tilde{J}_{\pm}, \tag{4.10.11}$$

$$\tilde{J}_{\pm} = -N_0 e\tilde{v}_{\pm}, \tag{4.10.12}$$

where

$$\tilde{v}_{\pm} = j\frac{e}{\gamma m_0\left(\omega - v_0 k_z \mp \omega_c\right)}\left(\tilde{E}_{\pm} \pm jv_0\tilde{B}_{\pm}\right), \tag{4.10.13}$$

and the double signs are in the same order.

From (4.10.10) ~ (4.10.13), we get for the dispersion relation for the transverse waves propagated along an unbounded beam,

$$\left[k_z^2 - \frac{\omega^2}{c^2} + \frac{\omega_p^2}{c^2}\frac{\omega - v_0 k_z}{\omega - v_0 k_z \mp \omega_c}\right]\tilde{E}_{\pm} = 0, \tag{4.10.14}$$

which is composed of the following two cases:

(i) $\tilde{E}_- = 0$ ($\tilde{E}_x = j\tilde{E}_y$);

$$k_z^2 - \frac{\omega^2}{c^2} + \frac{\omega_p^2}{c^2}\frac{\omega - v_0 k_z}{\omega - v_0 k_z - \omega_c} = 0, \tag{4.10.15}$$

(ii) $\tilde{E}_+ = 0$ ($\tilde{E}_x = -j\tilde{E}_y$);

$$k_z^2 - \frac{\omega^2}{c^2} + \frac{\omega_p^2}{c^2}\frac{\omega - v_0 k_z}{\omega - v_0 k_z + \omega_c} = 0. \tag{4.10.16}$$

For the case (i), which corresponds to the right-hand circularly polarized wave, the x component of electric field gains phase by $\pi/2$ as compared with the y component of electric field. On the other hand, for the case (ii), which corresponds to the left-hand circularly polarized wave, the x component of electric field loses phase by $\pi/2$ as compared with the y component of electric field. Thus the transverse waves propagated along a finitely magnetized unbounded beam are found to consist of right-hand and left-hand circularly polarized waves. For the special case where the applied magnetostatic field reduces to zero or infinity, the dispersion relation

(4.10.15) and (4.10.16) coincide with each other, when right-hand and left-hand circularly polarized waves are combined to make a linearly polarized wave.

For the case where no magnetostatic field is applied on the beam, or for the case of $\omega_c = 0$, the dispersion relations (4.10.15) and (4.10.16) reduce to

$$k_z^2 - \frac{\omega^2}{c^2} + \frac{\omega_p^2}{c^2} = 0, \tag{4.10.17}$$

which is the same as the dispersion relation for electromagnetic waves in a stationary electron plasma.

On the other hand, for the case where an infinite magnetostatic field is applied on the beam, or for the case of $\omega_c \rightarrow \infty$, we get from (4.10.15) and (4.10.16),

$$k_z^2 - \frac{\omega^2}{c^2} = 0, \tag{4.10.18}$$

which is the same as the dispersion relation for electromagnetic waves in vacuum. This means that an unbounded electron beam does not affect the propagation of electromagnetic waves along it when an infinite magnetostatic field is applied on it. For this case, electrons constituting the electron beam can move only in the direction of applied static magnetic field or in the direction of beam flow. Then, we have no transverse components of current density which affect the propagation of electromagnetic waves along the electron beam.

The transverse waves described by the dispersion relations (4.10.15) and (4.10.16) are composed of two kinds of waves, namely, electromagnetic waves and electron cyclotron waves. First, we show in Fig. 4.2 the dispersion curves for electromagnetic waves. These dispersion curves asymptotically approach the lines $\omega = \pm ck_z$ as ω tends to infinity. This means that the phase velocity and the group velocity of electromagnetic waves approach the vacuum velocity of light, c as ω tends to infinity. In addition, in Fig. 4.2, the dispersion curve 1 corresponds to a right-hand circularly polarized wave, and the dispersion curve 2 expresses a left-

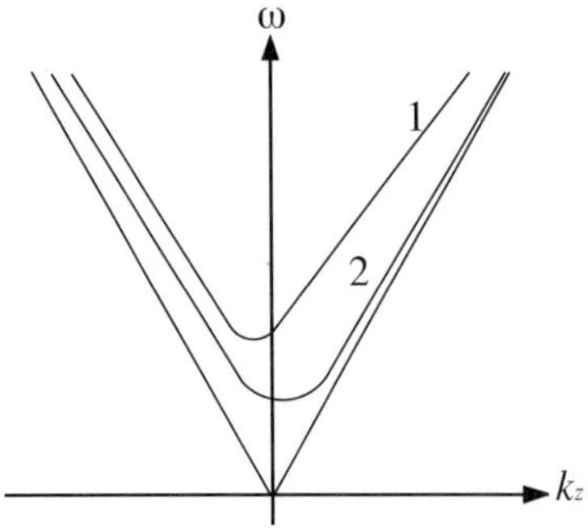

Fig. 4.2. Dispersion relation for electromagnetic waves in an unbounded beam

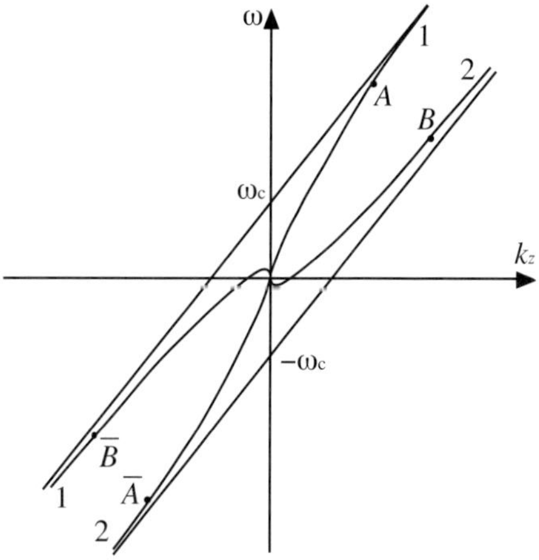

Fig. 4.3. Dispersion relations for electron-cyclotron waves in an unbounded beam

hand circularly polarized wave. In general, for the case where a finite magnetostatic field is applied on an unbounded beam, the dispersion curves for electromagnetic waves are separated into two branches, which degenerate into one branch (4.10.17) or (4.10.18) according as $\omega_c = 0$ or $\omega_c \to \infty$.

On the other hand, the dispersion curves for electron cyclotron waves are shown in Fig. 4.3. As seen from the figure, these dispersion curves asymptotically approach the lines $\omega - v_0 k_z \mp \omega_c = 0$ as $|\omega| \to \infty$. From (4.10.15) and (4.10.16), we find that these asymptotes represent the dispersion relations for electron cyclotron waves propagated in a sufficiently thin unbounded beam for which $\omega_p \to 0$. From these asymptotes, we find that the phase velocity and the group velocity approach the drift velocity of the electron beam, v_0 as $|\omega| \to \infty$. In Fig. 4.3, the part of $\omega > 0$ on the curve 1 and the part of $\omega < 0$ on the curve 2 represent right-hand circularly polarized waves, while the part of $\omega < 0$ on the curve 1 and the part of $\omega > 0$ on the curve 2 correspond to left-hand circularly polarized waves. In addition, for the case of $\omega_c = 0$ or $\omega_c \to \infty$, electron cyclotron waves disappear, and only electromagnetic waves for which the dispersion relation is given by (4.10.17) or (4.10.18) can be propagated.

In the analysis of electron beams, as shown in Fig. 4.3, we sometimes treat waves in the domain of $\omega < 0$ on dispersion curves. Let us consider a particular point (ω, k_z) in the domain of $\omega < 0$ on the dispersion curve. The wave corresponding to this point physically represents the same wave as that corresponding to the point $(-\omega, -k_z)$, which is symmetrical with the former point with respect to the origin. For example, in Fig. 4.3, the wave at the point $\overline{A}$ expresses the same wave as that at the point A, which is symmetrical with the point $\overline{A}$ with respect to the origin. In fact, the waves at the points $\overline{A}$ and A have the same values for the phase

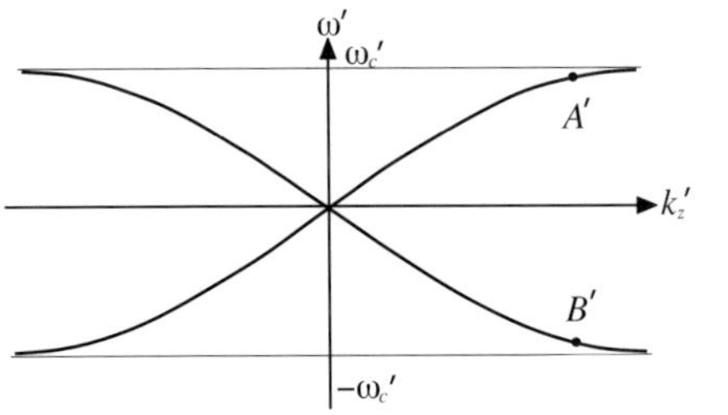

Fig. 4.4. Dispersion relations for electron-cyclotron waves in the rest frame of the electron beam

velocity and the group velocity. The above comments hold for the points $\overline{B}$ and B in Fig. 4.3 as well.

Next, in order to investigate the properties of electron cyclotron waves in more detail, we show in Fig. 4.4 the dispersion relation for these waves in the rest frame of the electron beam. The dispersion curves in this figure are the same as those for electron cyclotron waves in a stationary electron plasma. The dispersion relations in Figs. 4.3 and 4.4 can be transformed from one to the other through the Lorentz transformations for frequency and wave numbers given by (2.5.5). In Fig. 4.4, we show the points A' and B' corresponding to the points A and B in Fig. 4.3. The point A' represents the wave propagated in the positive z' direction, while the point B' corresponds to the wave propagated in the negative z' direction. On the other hand, the point A in Fig. 4.3 corresponding to the point A' in Fig. 4.4 represents the wave propagated in the positive z direction with the phase velocity greater than the drift velocity of the electron beam, while the point B corresponding to the point B' expresses the wave propagated in the positive z direction with the phase velocity less than the drift velocity of the electron beam.

4.10.2 Space-Charge Waves (Electron Plasma Waves)

For longitudinal waves propagated along an unbounded electron beam, we have the following basic equations from (4.2.7) and (4.2.8):

$$j\omega\varepsilon_0\tilde{E}_z + \tilde{J}_z = 0\,, \tag{4.10.19}$$

$$-jk_z\tilde{E}_z = \frac{\tilde{\rho}}{\varepsilon_0}\,, \tag{4.10.20}$$

where $\tilde{J}_z$ and $\tilde{\rho}$ are expressed from (4.2.4) and (4.2.2) as

$$\tilde{J}_z = -N_0 e\tilde{v}_z - \tilde{n}ev_0\,, \tag{4.10.21}$$

$$\tilde{\rho} = -\tilde{n}e\ . \tag{4.10.22}$$

In addition, from the relativistic equation of motion for the electron, we get

$$\tilde{v}_z = j\frac{e}{\gamma^3 m_0(\omega - v_0 k_z)}\tilde{E}_z. \tag{4.10.23}$$

Furthermore, from the equation of continuity (4.2.13) and (4.10.23), we have

$$\tilde{n} = \frac{N_0 k_z}{\omega - v_0 k_z}\tilde{v}_z = j\frac{N_0 e k_z}{\gamma^3 m_0(\omega - v_0 k_z)^2}\tilde{E}_z. \tag{4.10.24}$$

To find the dispersion relation for longitudinal waves, we do not need all the equations (4.10.19) ~ (4.10.24). Either of (4.10.20) with (4.10.22) or (4.10.24) is required in addition to the other equations. In the following analysis, we use (4.10.24) instead of (4.10.20) with (4.10.22). First, we insert (4.10.23) and (4.10.24) in (4.10.21) to express $\tilde{J}_z$ in terms of $\tilde{E}_z$. Then, substituting $\tilde{J}_z$ in (4.10.19), we obtain

$$j\omega\varepsilon_0\left[1 - \frac{\omega_p^2}{\gamma^2(\omega - v_0 k_z)^2}\right]\tilde{E}_z = 0. \tag{4.10.25}$$

Since $\tilde{E}_z$ does not vanish for longitudinal waves, we get, from (4.10.25), the following relation:

$$1 - \frac{\omega_p^2}{\gamma^2(\omega - v_0 k_z)^2} = 0, \tag{4.10.26}$$

which is rewritten as

$$\omega - v_0 k_z = \pm\frac{\omega_p}{\gamma}. \tag{4.10.27}$$

The relations (4.10.26) and (4.10.27) can also be found by using (4.10.20) with (4.10.22) instead of (4.10.24).

Equation (4.10.26) or (4.10.27) represents the dispersion relation for the longitudinal wave propagated along an unbounded electron beam, which is referred to as a space-charge wave or an electron plasma wave. The space-charge wave in the electron beam corresponds to a longitudinal oscillation of electrons in it or an electron plasma oscillation carried by its drift. From the above discussion, we find that the longitudinal wave in an unbounded electron beam is not affected by a magnetostatic field applied on it.

The dispersion relations for the space-charge wave are given by two straight lines, which are parallel to each other, as shown in Fig. 4.5. Of these two dispersion curves, one represents a wave with the phase velocity greater than the drift velocity of the electron beam, and the other corresponds to a wave with the phase velocity less than it. In Fig. 4.5, the points A and $\overline{A}$ express a fast space-charge wave, while the points B and $\overline{B}$ correspond to a slow space-charge wave. Thus we have two

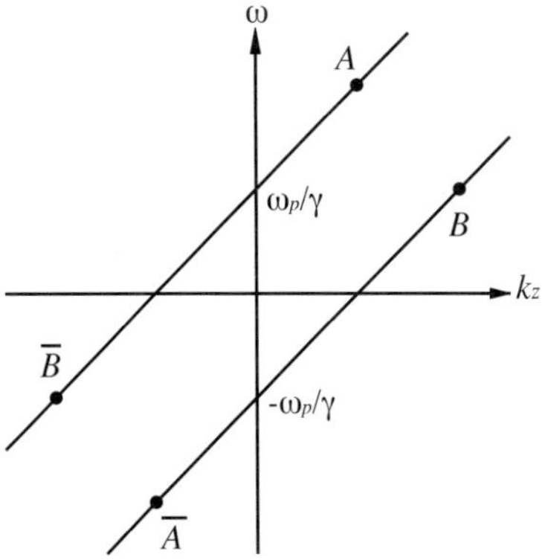

Fig. 4.5. Dispersion relations for space-charge waves in an unbounded beam

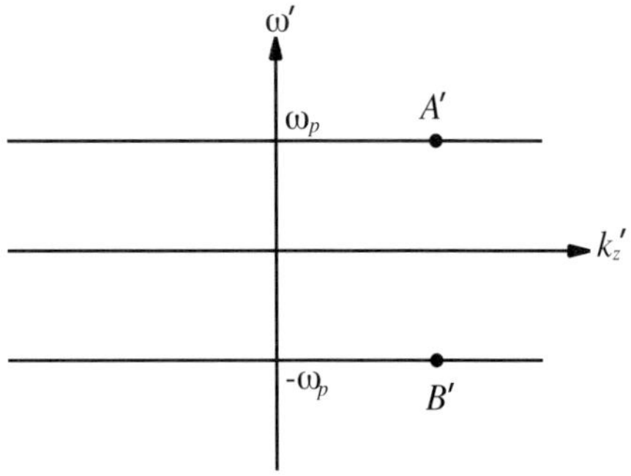

Fig. 4.6. Dispersion relations for space-charge waves in the rest frame of the electron beam

kinds of space-charge waves with different phase velocities propagated in the electron beam, the group velocity of which is equal to the drift velocity of the electron beam, v_0.

In Fig. 4.6, we show the dispersion relations for the space-charge wave in the rest frame of the electron beam, which correspond to those in the laboratory frame as shown in Fig. 4.5. The points A' and B' in Fig. 4.6 correspond to the points A and B in Fig. 4.5. The dispersion relations in Figs. 4.5 and 4.6 are related to each other through the Lorentz transformations for frequency and wave numbers as is the case with Figs. 4.3 and 4.4 for electron cyclotron waves. The points A' and B' in Fig. 4.6 represent waves propagated in the positive and negative z' directions, respectively. The dispersion relations for these waves are given by two straight lines parallel to the k_z'-axis. Thus the group velocity of these waves reduces to zero, which means that no energy is carried by them. In other words, in the rest frame of the electron beam, the space-charge waves are reduced to electron plasma oscillations. If we observe the waves represented by the points A' and B' from an inertial system, which is receding in the negative z' direction with the velocity greater than the phase velocity at the point B', we see a fast space-charge wave (point A) corresponding to the wave at the point A', and a slow space-charge wave (point B) corresponding to the wave at the point B'.

4.10.3 Energy Relations

The energy relations for small-signal fields propagated in a relativistic electron beam have been considered in detail in Sect. 4.5 on the basis of the polarization current model for the electron beam. In this section, we are discussing small-signal fields propagated in an unbounded beam with the aid of the convection current model. Thus we describe energy relations for small-signal fields from the standpoint of the latter model. From the basic field equations for the convection current model, (4.2.6) ~ (4.2.9), we get the following identity:

$$\nabla\cdot\left(\boldsymbol{E}\times\frac{1}{\mu_0}\boldsymbol{B}\right)=-\frac{\partial}{\partial t}\left(\frac{1}{2}\varepsilon_0\boldsymbol{E}\cdot\boldsymbol{E}+\frac{1}{2}\frac{1}{\mu_0}\boldsymbol{B}\cdot\boldsymbol{B}\right)-\boldsymbol{E}\cdot\boldsymbol{J}. \tag{4.10.28}$$

In the same manner as in Sect. 4.5, we consider quasi-sinusoidal fields with amplitudes varying slowly in terms of z and t, and average (4.10.28) temporally over one period of the quasi-sinusoidal fields. Then, we get the law of energy conservation for small-signal fields in the convection current model.

In order to find the time-average of (4.10.28), we need the relation between the Fourier components for the current density $\boldsymbol{J}$ and the electric field $\boldsymbol{E}$. This relation for transverse waves are found from (4.10.2) ~ (4.10.6) as

$$\tilde{\boldsymbol{J}}_t=\hat{\sigma}_t\cdot\tilde{\boldsymbol{E}}_t, \tag{4.10.29}$$

with

$$\hat{\sigma}_t=\frac{\gamma(\omega-v_0k_z)}{\omega}\begin{bmatrix}\sigma_{11} & \sigma_{12}\\ -\sigma_{12} & \sigma_{11}\end{bmatrix}, \tag{4.10.30}$$

where $\tilde{\boldsymbol{J}}_t$ and $\tilde{\boldsymbol{E}}_t$ are the transverse Fourier components for the current density and the electric field, σ_{11} and σ_{12} being given in (4.3.13).

On the other hand, for longitudinal waves, from (4.10.21), (4.10.23), and (4.10.24), we get

$$\tilde{J}_z=\sigma_l\tilde{E}_z, \tag{4.10.31}$$

where

$$\sigma_l=-j\frac{\varepsilon_0\omega_p^2\omega}{\gamma^2(\omega-v_0k_z)^2}. \tag{4.10.32}$$

Averaging (4.10.28) temporally over one period of quasi-sinusoidal fields, with the aid of (4.10.29) or (4.10.31), we find the same form of conservation law as (4.5.7). First, for transverse waves, the time-average values for the power flow density and the energy density, $\langle\boldsymbol{S}\rangle$ and $\langle w\rangle$ are given by

$$\langle\boldsymbol{S}\rangle=\frac{1}{2}\mathrm{Re}\left(\tilde{\boldsymbol{E}}\times\frac{1}{\mu_0}\tilde{\boldsymbol{B}}^*\right)+\boldsymbol{i}_z\left(\frac{1}{4}j\tilde{\boldsymbol{E}}^*\cdot\frac{\partial\hat{\sigma}_t}{\partial k_z}\cdot\tilde{\boldsymbol{E}}\right), \tag{4.10.33}$$

$$\langle w\rangle=\frac{1}{4}\varepsilon_0\tilde{\boldsymbol{E}}\cdot\tilde{\boldsymbol{E}}^*+\frac{1}{4}\frac{1}{\mu_0}\tilde{\boldsymbol{B}}\cdot\tilde{\boldsymbol{B}}^*-\frac{1}{4}j\tilde{\boldsymbol{E}}^*\cdot\frac{\partial\hat{\sigma}_t}{\partial\omega}\cdot\tilde{\boldsymbol{E}}, \tag{4.10.34}$$

in which (4.10.30) is inserted to obtain

$$\langle \boldsymbol{S} \rangle = \boldsymbol{i}_z \langle S \rangle, \ \langle S \rangle = \frac{k_z}{\omega \mu_0} \left| \tilde{E}_x \right|^2 \pm v_0 \left[\frac{1}{2} \varepsilon_0 \frac{\omega_p^2 \omega_c}{\omega \left(\omega - v_0 k_z \mp \omega_c \right)^2} \right] \left| \tilde{E}_x \right|^2, \tag{4.10.35}$$

$$\langle w \rangle = \varepsilon_0 \left| \tilde{E}_x \right|^2 \pm \frac{1}{2} \varepsilon_0 \frac{\omega_p^2 \omega_c}{\omega \left(\omega - v_0 k_z \mp \omega_c \right)^2} \left| \tilde{E}_x \right|^2, \tag{4.10.36}$$

where double signs are in the same order. For $\omega > 0$, the upper sign represents a right-hand circularly polarized wave, while the lower sign corresponds to a left-hand circularly polarized wave. On the other hand, for $\omega < 0$, the above relations are replaced with each other. For electromagnetic waves, the first term of (4.10.36) predominates, whereas the second term is predominant for electron cyclotron waves. Hence electromagnetic waves and electron cyclotron wave with right-hand circular polarization are positive-energy waves, while an electron cyclotron wave with left-hand circular polarization is a negative-energy wave. In addition, the power flow density and the energy density diverge when $\omega - v_0 k_z \mp \omega_c = 0$, which corresponds to the electron cyclotron resonance.

On the other hand, the power flow density and the energy density for longitudinal waves are found to be given by

$$\langle S \rangle = \frac{1}{4} j \frac{\partial \sigma_l}{\partial k_z} \tilde{\boldsymbol{E}} \cdot \tilde{\boldsymbol{E}}^*, \tag{4.10.37}$$

$$\langle w \rangle = \frac{1}{4} \varepsilon_0 \tilde{\boldsymbol{E}} \cdot \tilde{\boldsymbol{E}}^* - \frac{1}{4} j \frac{\partial \sigma_l}{\partial \omega} \tilde{\boldsymbol{E}} \cdot \tilde{\boldsymbol{E}}^*. \tag{4.10.38}$$

Inserting (4.10.32) in (4.10.37) and (4.10.38), and using the dispersion relation for longitudinal waves, (4.10.27), we get

$$\langle S \rangle = \pm \langle w \rangle v_0,$$

$$\langle w \rangle = \frac{1}{2} \frac{\omega}{\omega - v_0 k_z} \varepsilon_0 \left| \tilde{E}_z \right|^2 = \pm \frac{1}{2} \gamma \frac{\omega}{\omega_p} \varepsilon_0 \left| \tilde{E}_z \right|^2, \tag{4.10.39}$$

where the upper sign represents a fast space-charge wave and the lower sign corresponds to a slow space-charge wave. In other words, a fast space-charge wave is a positive-energy wave, while a slow space-charge wave is a negative-energy wave.

In Sect. 4.7, a simple physical meaning was given to positive-energy and negative-energy waves. Here, let us consider their physical meaning in a more specific manner for the case of space-charge waves. For this purpose, we investigate how the energy density is changed when small-signal longitudinal fields are excited in an unbounded electron beam. First, the variation in kinetic energy per unit volume of the electron beam, namely, the variation in kinetic energy density, ΔK can be calculated as

$$\Delta K = \left(N_0 + n\right)\left(\frac{1}{\sqrt{1 - \left(\frac{v_0 + v_z}{c}\right)}} - 1\right) m_0 c^2$$

$$- N_0 \left(\frac{1}{\sqrt{1 - \left(\frac{v_0}{c}\right)^2}} - 1\right) m_0 c^2 . \tag{4.10.40}$$

Expanding the first term on the right-hand side of (4.10.40) in a power series with respect to small-signal fields, and leaving the result to their second-order terms, we get

$$\frac{\Delta K}{m_0 c^2} = \gamma^3 \frac{N_0 v_0}{c^2} v_z + (\gamma - 1) n + \gamma^3 \frac{v_0}{c^2} n v_z + \frac{1}{2} \gamma^3 N_0 \left(\frac{v_z}{c}\right)^2 . \tag{4.10.41}$$

Averaging the above equation temporally over one period of quasi-sinusoidal small-signal fields, we obtain from the third and fourth terms on the right-hand side,

$$\frac{\langle \Delta K \rangle}{m_0 c^2} = \frac{1}{2} \gamma^3 \frac{v_0}{c^2} \mathrm{Re}\left(\tilde{n} \tilde{v}_z^*\right) + \frac{1}{4} \gamma^3 \frac{N_0}{c^2} \tilde{v}_z \tilde{v}_z^* . \tag{4.10.42}$$

Represent the right-hand side of the above equation in terms of the z component of electric field, with the aid of (4.10.23) and (4.10.24), and rewrite the resultant equation, using the dispersion relation for space-charge waves, (4.10.27). Then, we have the final result for $<\Delta K>$

$$\langle \Delta K \rangle = \frac{1}{4} \frac{\omega + v_0 k_z}{\omega - v_0 k_z} \varepsilon_0 \tilde{E}_z \tilde{E}_z^* . \tag{4.10.43}$$

On the other hand, when small-signal fields are excited in an unbounded electron beam, we find for the time-average value of the variation in electric energy density, $<\Delta w_e>$,

$$\langle \Delta w_e \rangle = \frac{1}{4} \varepsilon_0 \tilde{E}_z \tilde{E}_z^* . \tag{4.10.44}$$

From the above discussion, when small-signal longitudinal fields are excited along an unbounded beam, the time-average value of the total variation in energy density, $<\Delta U>$ is expressed as the sum of $<\Delta K>$ and $<\Delta w_e>$,

$$<\Delta U> = <\Delta K> + <\Delta w_e>. \tag{4.10.45}$$

Add (4.10.43) and (4.10.44), and rearrange the resultant equation with the aid of the dispersion relation for space-charge waves, (4.10.27). Then, we come to the final result for $<\Delta U>$:

$$\langle \Delta U \rangle = \frac{1}{2} \frac{\omega}{\omega - v_0 k_z} \varepsilon_0 \left| \tilde{E}_z \right|^2 = \pm \frac{1}{2} \gamma \frac{\omega}{\omega_p} \varepsilon_0 \left| \tilde{E}_z \right|^2 . \qquad (4.10.46)$$

We find that Eq. (4.10.46) is equal to the energy density for space-charge waves given by (4.10.39). In most cases, electron beams are used under the condition $\omega_p << \omega$. For these cases, as is evident from (4.10.46) and (4.10.44), it follows that $\langle \Delta U \rangle >> \langle \Delta w_e \rangle$, which means that most of the energy carried by space-charge waves are kinetic energy.

In conclusion, when small-signal fields are excited in the electron beam, the time-average value of the total variation in energy density becomes positive for positive-energy waves and negative for negative-energy waves. This is the physical meaning of positive-energy and negative-energy waves.

5. Stimulated Cherenkov Effect

5.1 Generation of Growing Waves by Stimulated Cherenkov Effect

We have clarified in Sect. 4.7 that a relativistic electron beam can support waves carrying negative energy. We can make negative-energy waves interact with positive-energy waves to generate or amplify high-power electromagnetic waves of short wavelengths. A typical example of generators or amplifiers based on this principle is a traveling wave tube in the microwave region. In the traveling wave tube, a slow space-charge wave propagated along an electron beam (negative-energy wave) is made to be coupled with an electromagnetic wave propagated along a helical conductor (positive-energy wave) to generate or amplify electromagnetic waves in the microwave frequency range. As described in Chap. 4, the negative-energy wave in an electron beam is propagated with a velocity nearly equal to its drift velocity. Hence, in order to get efficient coupling between the electron beam and the electromagnetic wave, we must reduce the phase velocity of the electromagnetic wave so that it becomes approximately equal to the drift velocity of the electron beam. Wave-guiding structures to decrease the phase velocity of the electromagnetic wave are called slow-wave structures. In the traveling wave tube, a helical conductor serves as a slow wave structure. Waveguides with periodic structures or dielectric-loaded waveguides are other examples of slow wave structures.

In this chapter, we discuss the active coupling of a relativistic electron beam and electromagnetic waves propagated along a dielectric waveguide from two alternative approaches: those based on the fluid model and the single-particle model of the electron beam. In the former approach based on the fluid model of the electron beam, we consider the coupling of a negative-energy space-charge wave propagated along a relativistic electron beam and an electromagnetic wave propagated along a dielectric waveguide. Among various frequency component waves generated by the Cherenkov radiation, only those waves which satisfy appropriate dispersion relations for the waveguide and phase-matching conditions with the space-charge wave can survive and continuously get energy from the electron beam, growing as they propagate along with it. This approach may also be called the collective approach because the electron beam is treated collectively as a continuous medium in this approach. On the other hand, in the latter approach based on the single-particle model of the electron beam, we investigate the interaction between each of electrons composing a relativistic electron beam and an electromagnetic wave propagated along a dielectric waveguide. Those electrons traveling with velocities nearly equal to the phase velocity of the electromagnetic wave strongly interact with the electromagnetic wave. Then, if the average velocity of electrons constituting a relativistic electron beam is greater than the phase velocity of the electromagnetic wave, the electromagnetic wave can get net energy from the electron beam and, as a result, the former grows at the expense of the kinetic energy of the latter. In either of the above two approaches, the amplified energy of the electromagnetic wave is supplied from the kinetic energy of the electron beam through the continuous process of the stimulated Cherenkov radiation or the stimulated Cherenkov effect. In the Cherenkov free-electron laser, high-power radiation in the millimeter to submillimeter wave regions can be generated through the stimulated Cherenkov effect.

5.2 Field Expressions in the Relativistic Electron Beam

Let us consider a two-dimensional model composed of a dielectric-coated conducting plane of infinite extent and a semi-infinite relativistic electron beam separated from the former by a narrow vacuum gap, as shown in Fig. 5.1. For simplicity, we assume that the relativistic electron beam is not magnetized and uniformly drifting in the z direction. On these assumptions and on the basis of the fluid model of the electron beam, we discuss the coupling between the electromagnetic wave and the space-charge wave propagated along the system shown in Fig. 5.1, and clarify that the system can support growing waves of which amplitudes increase exponentially in the direction of the beam flow.

To prepare for the following discussion, let us try to find the general representations for field components of waves propagated in a two-dimensional relativistic electron beam. For the space-charge wave, the electric field component in the z direction, namely, in the direction of the beam flow plays an important role. Hence we consider the propagation of TM space-charge waves, which have a non-

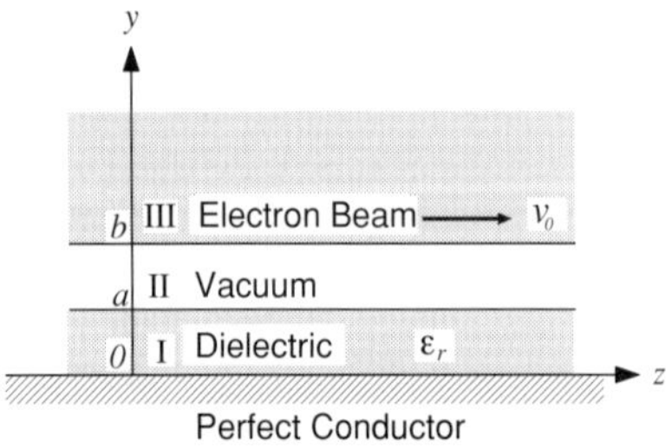

Fig. 5.1. Geometry of the problem

zero z component of electric field. Let an arbitrary component of small-signal fields F be represented in complex form as

$$F = \tilde{F}(y)\exp[j(\omega t - k_z z)]. \tag{5.2.1}$$

Then, from the Maxwell equations for small-signal fields, (4.2.6) ~ (4.2.9), we get for TM waves,

$$\frac{\partial \tilde{E}_z}{\partial y} + jk_z\tilde{E}_y = -j\omega\tilde{B}_x, \tag{5.2.2}$$

$$-jk_z\tilde{B}_x = j\frac{\omega}{c^2}\tilde{E}_y - \mu_0 e N_0 \tilde{v}_y, \tag{5.2.3}$$

$$-\frac{\partial \tilde{B}_x}{\partial y} = j\frac{\omega}{c^2}\tilde{E}_z - \mu_0 e(\tilde{n}v_0 + N_0\tilde{v}_z), \tag{5.2.4}$$

$$\frac{\partial \tilde{E}_y}{\partial y} - jk_z\tilde{E}_z = -\frac{e}{\varepsilon_0}\tilde{n}, \tag{5.2.5}$$

where $-e$ is the electronic charge, and v_0 is the drift velocity of the electron beam, $\tilde{v}_y$ and $\tilde{v}_z$ denoting the y and z components of the small-signal velocity. In addition, N_0 and $\tilde{n}$ express the average number density and the small-signal number density of electrons, respectively.

The small-signal quantities $\tilde{v}_y$ and $\tilde{v}_z$ are obtained from the relativistic equation of motion for the electron, while the small-signal quantity $\tilde{n}$ is found from the equation of continuity for the flow of electrons. First, for a nonmagnetized relativistic electron beam, we get from (4.2.10),

$$\tilde{v}_y = j\frac{e}{\gamma m_0(\omega - v_0 k_z)}\left(\tilde{E}_y + v_0\tilde{B}_x\right), \tag{5.2.6}$$

$$\tilde{v}_z = j\frac{e}{\gamma^3 m_0(\omega - v_0 k_z)}\tilde{E}_z, \tag{5.2.7}$$

where m_0 is the rest mass of the electron.

For $\tilde{n}$, we find from (4.2.13),

$$\tilde{n} = j\frac{N_0}{\omega - v_0 k_z}\left(\frac{\partial \tilde{v}_y}{\partial y} - jk_z\tilde{v}_z\right). \tag{5.2.8}$$

Substituting (5.2.6) and (5.2.7) in (5.2.8) to represent $\tilde{n}$ in terms of electromagnetic field components, we have

$$\tilde{n} = j\frac{N_0 e k_z}{\gamma^3 m_0(\omega - v_0 k_z)^2}\tilde{E}_z - \frac{N_0 e}{\gamma m_0(\omega - v_0 k_z)^2}\frac{\partial}{\partial y}\left(\tilde{E}_y + v_0 \tilde{B}_x\right). \tag{5.2.9}$$

Inserting (5.2.6) in (5.2.3) yields the relation between $\tilde{B}_x$ and $\tilde{E}_y$,

$$\tilde{B}_x = -\frac{1}{c^2}\frac{\omega_p^2 - \omega(\omega - v_0 k_z)}{(\beta / c)\omega_p^2 - k_z(\omega - v_0 k_z)}\tilde{E}_y. \tag{5.2.10}$$

Using (5.2.10) in (5.2.9), and then inserting the resultant $\tilde{n}$ in (5.2.5), we find

$$\frac{\partial \tilde{E}_y}{\partial y} = -j\frac{(\beta / c)\omega_p^2 - k_z(\omega - v_0 k_z)}{\omega - v_0 k_z}\tilde{E}_z. \tag{5.2.11}$$

Similarly, from (5.2.2) and (5.2.10), we get

$$\frac{\partial \tilde{E}_z}{\partial y} = jk_y^2\frac{\omega - v_0 k_z}{(\beta / c)\omega_p^2 - k_z(\omega - v_0 k_z)}\tilde{E}_y, \tag{5.2.12}$$

where

$$k_y^2 = k_z^2 - \left(\frac{\omega}{c}\right)^2 + \left(\frac{\omega_p}{c}\right)^2. \tag{5.2.13}$$

Differentiating both sides of (5.2.12) with respect to y, and substituting (5.2.11) in the resultant equation, we obtain the differential equation for $\tilde{E}_z$,

$$\frac{\partial^2 \tilde{E}_z}{\partial y^2} - k_y^2 \tilde{E}_z = 0. \tag{5.2.14}$$

We are now ready to get the field components $\tilde{E}_z$, $\tilde{E}_y$, and $\tilde{B}_x$ for TM waves propagated along the two-dimensional relativistic electron beam. First, we can get the field component $\tilde{E}_z$ by solving (5.2.14). Then, using $\tilde{E}_z$ in (5.2.12), we obtain the field component $\tilde{E}_y$. Finally, we have the field component $\tilde{B}_x$ from (5.2.10). In the model shown in Fig. 5.1, the fields in the relativistic electron beam must vanish as $y \to \infty$. Hence we get the condition $k_y^2 > 0$. Thus we find for the field components of TM waves propagated along the two-dimensional relativistic electron beam under consideration,

$$\tilde{E}_z = Ce^{-k_y y}, \tag{5.2.15}$$

$$\tilde{E}_y = j\frac{(\beta / c)\omega_p^2 - k_z(\omega - v_0 k_z)}{k_y(\omega - v_0 k_z)}Ce^{-k_y y}, \tag{5.2.16}$$

$$\tilde{B}_x = -j\frac{\omega_p^2 - \omega(\omega - v_0 k_z)}{c^2 k_y(\omega - v_0 k_z)} C e^{-k_y y}, \tag{5.2.17}$$

where C is a constant amplitude to be determined later from the boundary conditions.

5.3 Field Expressions in the Dielectric and Vacuum Regions

The field components in the dielectric region can be obtained by setting $N_0 = \tilde{n} = 0$ and replacing c^2 with c^2/ε_r in (5.2.2) ~ (5.2.5), where ε_r denotes the relative permittivity of the dielectric. The z component of electric field in the dielectric region is found to satisfy

$$\frac{\partial^2 \tilde{E}_z}{\partial y^2} + p_y^2 \tilde{E}_z = 0, \tag{5.3.1}$$

with

$$p_y^2 = \varepsilon_r \left(\frac{\omega}{c}\right)^2 - k_z^2, \tag{5.3.2}$$

where we have the condition $p_y^2 > 0$ since the phase velocity in a dielectric with finite thickness must be greater than that in an unbounded dielectric. We first solve (5.3.1) with the condition $\tilde{E}_z = 0$ at $y = 0$. Then, with the aid of this solution, we find the field components $\tilde{E}_y$ and $\tilde{B}_x$. Thus we have for the field components in the dielectric region,

$$\tilde{E}_z = A \sin p_y y, \tag{5.3.3}$$

$$\tilde{E}_y = -j\frac{k_z}{p_y} A \cos p_y y, \tag{5.3.4}$$

$$\tilde{B}_x = j\frac{\omega \varepsilon_r}{c^2 p_y} A \cos p_y y, \tag{5.3.5}$$

where A is an unknown constant to be determined from the boundary conditions.

In the vacuum region, on the other hand, the field components can be found from (5.2.2) ~ (5.2.5) with $N_0 = \tilde{n} = 0$. The z component of electric field in this region is found to satisfy

$$\frac{\partial^2 \tilde{E}_z}{\partial y^2} - h_y^2 \tilde{E}_z = 0, \tag{5.3.6}$$

with

$$h_y^2 = k_z^2 - \left(\frac{\omega}{c}\right)^2, \qquad (5.3.7)$$

where we get the condition $h_y^2 > 0$ from the fact that the phase velocity of waves propagated in the system shown in Fig. 5.1 is less than the velocity of light in vacuum c. With the aid of the solution of (5.3.6), we can express the field components in the vacuum region as

$$\tilde{E}_z = D_1 \cosh h_y y + D_2 \sinh h_y y, \qquad (5.3.8)$$

$$\tilde{E}_y = j\frac{k_z}{h_y}(D_1 \sinh h_y y + D_2 \cosh h_y y), \qquad (5.3.9)$$

$$\tilde{B}_x = -j\frac{\omega}{c^2 h_y}(D_1 \sinh h_y y + D_2 \cosh h_y y), \qquad (5.3.10)$$

where D_1 and D_2 are unknown constants to be determined from the boundary conditions.

5.4 Dispersion Relation and Growth Rate

In order to clarify that the system shown in Fig. 5.1 can support waves growing in the direction of the flow of the relativistic electron beam, we first derive the dispersion relation for coupled electromagnetic and space-charge waves by imposing the boundary conditions at the surfaces of the dielectric and the relativistic electron beam on the field components of the coupled waves. Then, solving the dispersion relation for the wave number in the z direction k_z, we show that one of the solutions for k_z has a positive imaginary part, which corresponds to a growing wave.

The boundary conditions at the surface of the dielectric ($y = a$) are given by

$$\tilde{E}_{1z} - \tilde{E}_{2z} = 0, \qquad (5.4.1)$$

$$\tilde{B}_{1x} - \tilde{B}_{2x} = 0, \qquad (5.4.2)$$

where the permeability of the dielectric is assumed to be equal to μ_0, and the subscripts 1 and 2 attached to field components refer to the dielectric and vacuum regions I and II, respectively. On the other hand, the boundary conditions at the surface of the electron beam ($y = b$) are obtained from (4.4.1) and (4.4.6), being expressed as

$$\tilde{E}_{2z} - \tilde{E}_{3z} = 0, \qquad (5.4.3)$$

$$\tilde{B}_{2x} - \tilde{B}_{3x} + \frac{\beta}{c}(\tilde{E}_{2y} - \tilde{E}_{3y}) = 0, \tag{5.4.4}$$

where the subscript 3 refers to the beam region III.

Imposing the boundary conditions (5.4.1) ~ (5.4.4) on the field components in the dielectric, vacuum and beam regions obtained earlier, we have the following set of linear algebraic equations for unknowns A, D_1, D_2, and C:

$$\begin{bmatrix} \sin p_y a & -\cosh h_y a & -\sinh h_y a & 0 \\ \frac{\varepsilon_r}{p_y}\cos p_y a & \frac{1}{h_y}\sinh h_y a & \frac{1}{h_y}\cosh h_y a & 0 \\ 0 & \cosh h_y b & \sinh h_y b & -e^{-k_y b} \\ 0 & \frac{1}{h_y}\sinh h_y b & \frac{1}{h_y}\cosh h_y b & \frac{\varepsilon_p}{k_y}e^{-k_y b} \end{bmatrix} \begin{bmatrix} A \\ D_1 \\ D_2 \\ C \end{bmatrix} = 0, \tag{5.4.5}$$

where

$$\varepsilon_p = 1 - \frac{\omega_p^2}{\gamma^2(\omega - v_0 k_z)^2}, \tag{5.4.6}$$

ε_p being the equivalent relative permittivity of the relativistic electron beam in its rest frame. If a set of linear algebraic equations (5.4.5) is to have nontrivial solutions for unknowns A, D_1, D_2, and C, it is necessary that the determinant composed of their coefficients vanish. Hence we get the following dispersion relation for coupled electromagnetic and space-charge waves propagated along the two-dimensional system depicted in Fig. 5.1:

$$\varepsilon_p\left[\frac{p_y}{k_y}\tan p_y a - \varepsilon_r\frac{h_y}{k_y}\tanh h_y(b-a)\right] + \frac{p_y}{h_y}\tan p_y a \tanh h_y(b-a) - \varepsilon_r = 0, \tag{5.4.7}$$

Equation (5.4.7) represents the dispersion relation for waves produced by the coupling of the electromagnetic wave, which propagates along the dielectric-coated conducting plane in the absence of the electron beam, and the space-charge wave, which propagates along the relativistic electron beam in the absence of the dielectric.

In the subsequent discussion, let us first investigate the properties of the electromagnetic and space-charge waves in the absence of their coupling. The dispersion relation for the electromagnetic wave propagated in the absence of the electron beam can be obtained by setting $\omega_p = 0$ and $k_y = h_y$ in (5.4.7) with the result

$$\frac{p_y}{h_y}\tan p_y a - \varepsilon_r = 0, \tag{5.4.8}$$

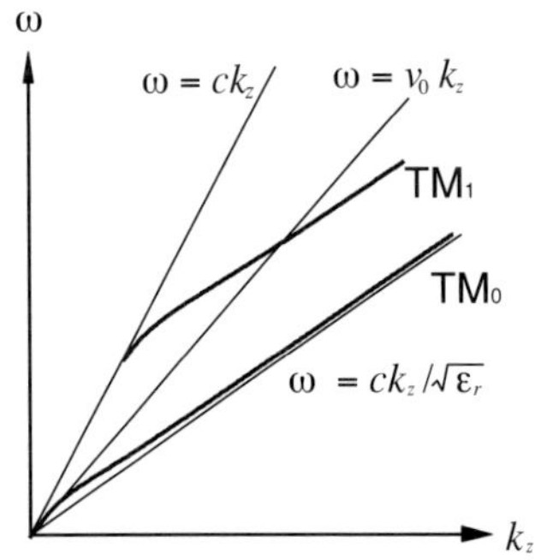

Fig. 5.2. Dispersion relations for electromagnetic waves propagated along a dielectric-coated conducting plane

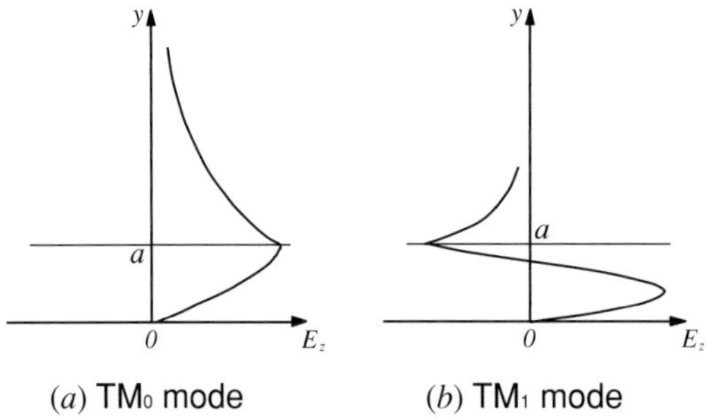

Fig. 5.3. Field distributions for TM_0 and TM_1 modes

which is the well-known dispersion relation for the surface wave propagated along a dielectric-coated conducting plane waveguide. In Fig. 5.2, we illustrate the dispersion curves for a couple of the lowest-order modes, namely, the TM_0 and TM_1 modes derived from (5.4.8). In addition, we show in Fig. 5.3 the distributions of the z component of electric field for these modes. As seen from Fig. 5.2, the dispersion curves obtained from (5.4.8) are located between the lines $\omega = ck_z$ and $\omega = ck_z/(\varepsilon_r)^{1/2}$ on the dispersion diagram. The dispersion curves in Fig. 5.2 asymptotically approach the latter line as ω tends to infinity, and become cutoff on the former line. The cutoff frequency ω_{cn} is given by

$$\frac{\omega_{cn}a}{c} = \frac{n\pi}{\sqrt{\varepsilon_r - 1}}, \quad (n = 0, 1, 2, \cdots). \tag{5.4.9}$$

On the other hand, the dispersion relation for the space-charge wave propagated in the absence of the dielectric can be found by setting $a = 0$ and $\varepsilon_r = 1$ in (5.4.7), taking the form

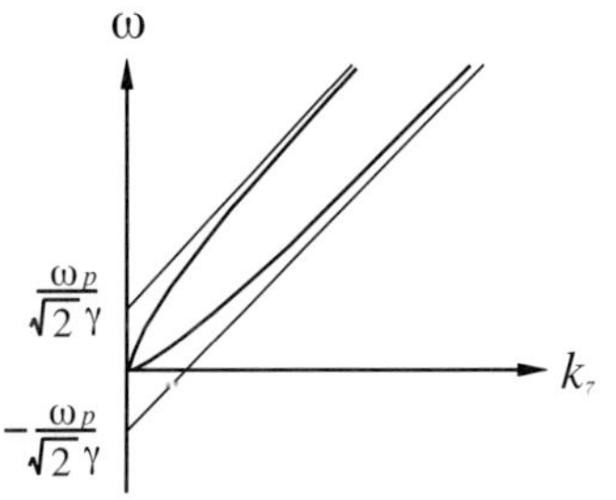

Fig. 5.4. Dispersion relations for space-charge waves propagated along a semi-infinite beam

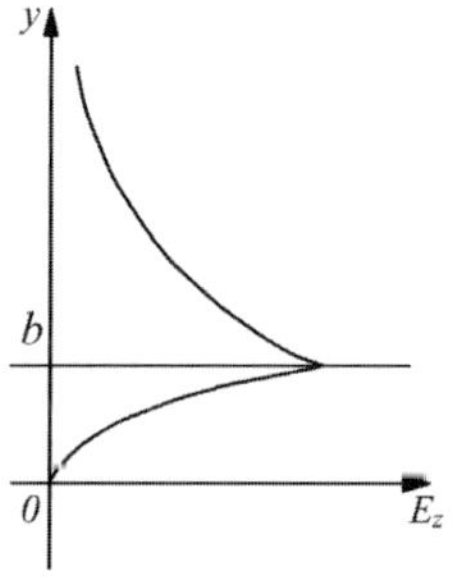

Fig. 5.5 Field distribution for space-charge waves propagated along a semi-infinite beam

$$\varepsilon_p \frac{h_y}{k_y} \tanh h_y b + 1 = 0. \tag{5.4.10}$$

The dispersion relation (5.4.10) is composed of two branches, as shown in Fig. 5.4. Of these two branches, the upper branch expresses the fast space-charge wave (positive-energy wave) while the lower branch corresponds to the slow space-charge wave (negative-energy wave). As ω tends to infinity, these dispersion curves asymptotically approach the lines

$$\omega - v_0 k_z = \pm \frac{1}{\sqrt{2}} \frac{\omega_p}{\gamma}. \tag{5.4.11}$$

The electric field distribution for the space-charge waves described by the dispersion relations (5.4.10), which is concentrated around the surface of the electron beam ($y = b$), is illustrated in Fig. 5.5.

Returning to the dispersion relation (5.4.7), let us consider the coupling of the electromagnetic wave and the space-charge wave. The electromagnetic and space-charge waves strongly interact with each other around the point where the phase velocities of these waves become equal, or around the intersection of the dispersion curves for them given by (5.4.8) and (5.4.10). There exist two intersections of the dispersion curves for the electromagnetic and space-charge waves. Of these two intersections, we get growing waves around one on the lower branch of the dispersion curves for the space-charge wave, which represents the negative-energy wave.

For simplicity, let us assume that the density of the electron beam is so low that the condition $\omega_p \ll \omega$ holds. Under this condition, we derive, from the dispersion relation (5.4.7), the growth rate for growing waves generated by the coupling of electromagnetic and space-charge waves. For a low-density electron beam, the dispersion relation (5.4.7) can be reduced, in the vicinity of the intersection of dispersion curves (5.4.8) and (5.4.10), to

$$e^{2h_y(b-a)}\left(\frac{p_y}{h_y}\tan p_y a-\varepsilon_r\right)(\omega-v_0k_z)^2=\varepsilon_r\left(\frac{\omega_p}{\gamma}\right)^2. \tag{5.4.12}$$

For a low-density electron beam, a set of ω and k_z to satisfy (5.4.7) can be considered to fall around the intersection of the dispersion curve (5.4.8) and the line $\omega = v_0k_z$. Then, let the values of ω and k_z at this intersection be ω_0 and k_{z0}, and let us try to find out the value of k_z to satisfy (5.4.12), with the value of ω fixed to ω_0. For this purpose, we express k_z around k_{z0} as

$$k_z = k_{z0} + \delta k_z, \tag{5.4.13}$$

where the condition $|\delta k_z| << k_{z0}$ is assumed to hold. Substituting (5.4.13) in (5.4.12), we get

$$(\delta k_z)^3=\varepsilon_r\left(\frac{\omega_p}{\gamma v_0}\right)^2\frac{e^{-2h_{y0}(b-a)}}{\dfrac{\partial}{\partial k_z}\left(\dfrac{p_y}{h_y}\tan p_y a\right)}, \tag{5.4.14}$$

where

$$h_{y0} = k_{z0}/\gamma. \tag{5.4.15}$$

In (5.4.14), note that the differentiation with respect to k_z is carried out at $\omega = \omega_0$ and $k_z = k_{z0}$. From (5.3.2) and (5.3.7), we find that the following relation holds:

$$\frac{\partial}{\partial k_z}\left(\frac{p_y}{h_y}\tan p_y a\right)<0, \tag{5.4.16}$$

at $\omega = \omega_0$ and $k_z = k_{z0}$. Hence we can rewrite (5.4.14) as

$$(\delta k_z)^3=-\varepsilon_r\left(\frac{\omega_p}{\gamma v_0}\right)^2\frac{e^{-2h_{y0}(b-a)}}{\left|\dfrac{\partial}{\partial k_z}\left(\dfrac{p_y}{h_y}\tan p_y a\right)\right|}, \tag{5.4.17}$$

from which we get

$$\delta k_z=\varepsilon_r^{1/3}\left(\frac{\omega_p}{\gamma v_0}\right)^{2/3}\frac{e^{-(2/3)h_{y0}(b-a)}}{\left|\dfrac{\partial}{\partial k_z}\left(\dfrac{p_y}{h_y}\tan p_y a\right)\right|^{1/3}}e^{j\frac{2m+1}{3}\pi},\quad (m=0,1,2). \tag{5.4.18}$$

From (5.4.18), we find that the dispersion relation (5.4.12) allows three values of k_z for one value of ω. Among them, one takes a real value, and the rest are complex quantities of which one is a complex conjugate to the other. For $m = 0$, in particular, the imaginary part of k_z becomes positive, giving growing waves. Then, the imaginary part of k_z, which is designated as α, corresponds to the growth rate of growing waves, the value of which is found from (5.4.18) as

$$\alpha = \frac{\sqrt{3}}{2}\varepsilon_r^{1/3}\left(\frac{\omega_p}{\gamma v_0}\right)^{2/3}\frac{e^{-(2/3)h_{y0}(b-a)}}{\left|\frac{\partial}{\partial k_z}\left(\frac{p_y}{h_y}\tan p_y a\right)\right|^{1/3}}. \tag{5.4.19}$$

The growth rate α means the exponential growth of waves per unit length. As seen from (5.4.19), the growth rate exponentially decreases as the beam-dielectric gap (b-a) increases. The reason for this fact can be explained physically as follows. From Figs. 5.3 and 5.5, we note that the field distributions for the electromagnetic and space-charge waves are concentrated around the dielectric and beam surfaces and exponentially reduced as receding from these surfaces. Hence, as the beam-dielectric gap increases, the interaction of electromagnetic and space-charge waves is rapidly weakened, leading to an exponential decrease in the growth rate.

For simplicity, we have treated a semi-infinite planar relativistic electron beam. However, in order to have the growth rate given by (5.4.19), a semi-infinite electron beam is not necessarily required. This is because the fields of the space-charge wave decay exponentially as receding from the beam surface, as stated in the foregoing discussion, and only a thin layer around the beam surface contributes to the interaction with the electromagnetic wave. Therefore, for the beam thickness over the skin depth of the electron beam, we can expect that almost the same value as (5.4.19) is safely obtained for the growth rate. By the skin depth of the electron beam, we mean the beam thickness at which the fields decay to $1/e$ of their values at the beam surface. Specifically, the skin depth of the electron beam, δ can be found from the relation

$$k_y\delta \approx h_y\delta = 1. \tag{5.4.20}$$

At the point of $\omega = \omega_0$ and $k_z = k_{z0}$, we have

$$h_y = k_{z0}/\gamma = \omega_0/\gamma v_0. \tag{5.4.21}$$

From (5.4.20) and (5.4.21), we get for the skin depth of the relativistic electron beam,

$$\delta = 1/h_y = \gamma v_0/\omega_0. \tag{5.4.22}$$

From the viewpoint of applications, the relation (5.4.22) is very important. The reason for it can be explained in the following manner. In order to get efficient coupling between the electron beam and the electromagnetic wave, the beam-dielectric gap must be kept as narrow as possible. However, this requirement seems

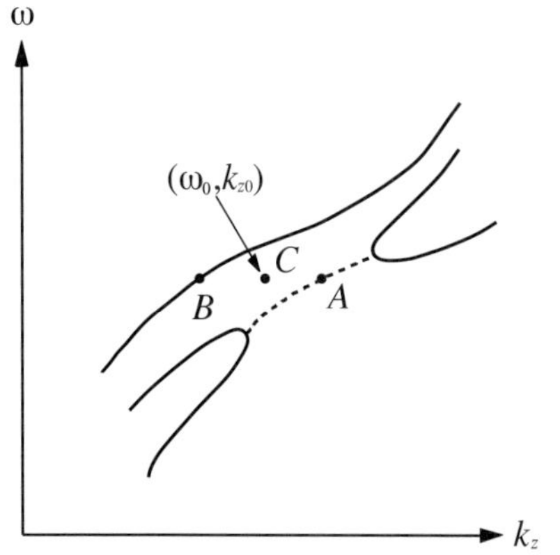

Fig. 5.6. Dispersion diagram in the vicinity of the coupling point

very difficult to clear for nonrelativistic electron beams. For relativistic electron beams, on the other hand, the condition for efficient beam-wave coupling becomes more easy to satisfy, because a wider beam-dielectric gap is allowed for the latter owing to the factor γ appearing in (5.4.22).

On the basis of the foregoing discussion, we can draw a rough picture of the dispersion relation (5.4.12) in the vicinity of the coupling point on the dispersion diagram, as shown in Fig. 5.6. In the figure, the point C denotes the intersection of the dispersion curve for the electromagnetic wave (5.4.8) and the line $\omega = v_0 k_z$. In addition, the point A corresponds to the wave numbers with $m = 0$ and $m = 2$ in (5.4.18), while the point B expresses the wave number for $m = 1$. We get growing waves on the broken line in the figure, and the growth rate becomes maximum at the point A while it drops to zero at both ends of the broken line. In Fig. 5.7, we give a numerical example of frequency dependence of the growth rate around the point A.

In this section, we have shown that growing waves are generated by coupling the electromagnetic wave and the slow space-charge wave propagated along the system composed of a dielectric-coated conducting plane, which serves as a slow-wave structure, and a semi-infinite relativistic electron beam, as shown in Fig. 5.1. Let us consider the physical reason why growing waves are obtained by the coupling of the electromagnetic wave and the slow space-charge wave. First, let a slow space-charge wave be launched in an electron beam. Then, the electron beam is decelerated and lose energy to an electromagnetic wave propagating with the same phase velocity as the slow space-charge wave, leading to the growth of the electromagnetic wave. On the other hand, as the amplitude of the electromagnetic wave grows, the electron beam is further decelerated by the amplified electric field of the former, which results in the growth of the slow space-charge wave. In the subsequent similar processes, a mechanism of positive feedback automatically works between the electromagnetic wave and the slow space-charge wave, leading to the

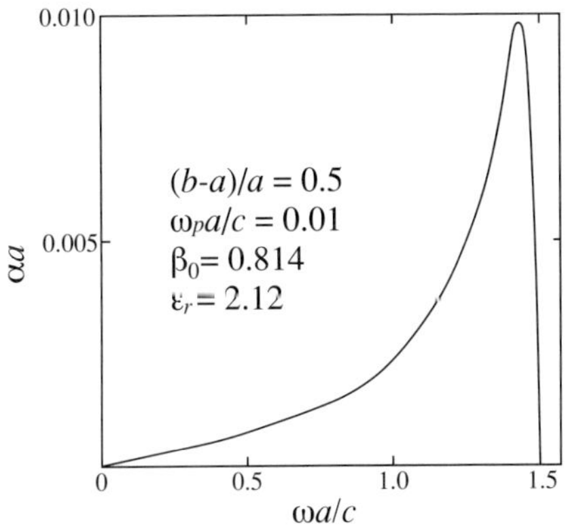

Fig. 5.7. Frequency dependence of the growth rate

growth of both waves. The growth of the slow space-charge wave, which is a negative-energy wave, corresponds to the decrease in the kinetic energy of the electron beam. Thus it follows that the growing energy of the electromagnetic wave is extracted from the kinetic energy of the electron beam.

For growing waves to be generated by the coupling of the electromagnetic wave and the space-charge wave, the drift velocity of the electron beam, v_0 must be greater than the phase velocity of the plane wave in the dielectric or the velocity of light in the dielectric, $c/(\varepsilon_r)^{1/2}$, as is evident from the dispersion relation shown in Fig. 5.2. This condition corresponds to that under which the Cherenkov radiation is produced in the dielectric. Hence the mechanism to generate short-wavelength electromagnetic waves by such a structure composed of a dielectric waveguide as shown in Fig. 5.1 is referred to as the stimulated Cherenkov effect, and the generator of the millimeter to submillimeter waves based on this mechanism is what is called the Cherenkov oscillator or the Cherenkov laser.

5.5 Power Transfer from the Electron Beam to the Electromagnetic Wave

The electromagnetic wave mode grows as it gets energy from the electron beam through the active coupling with the space-charge wave mode. Then, the drift velocity of the electron beam gradually decreases as it gives kinetic energy to the electromagnetic wave. Let us calculate the power transfer from the electron beam to the electromagnetic wave, taking into account the decrease in the drift velocity of the electron beam, on the basis of the coupled-mode theory [5.10]. For this purpose, denote the power carried by the electromagnetic wave mode as $P(z)$, and the power carried by the space-charge wave mode as $Q(z)$, assuming that $P(z) = P_0$ and $Q(z) = 0$ at $z = 0$. Then, the powers $P(z)$ and $Q(z)$ are expressed as

$$P(z) = P_0 \cosh^2 \alpha z, \tag{5.5.1}$$

$$Q(z) = -P_0 \sinh^2 \alpha z, \tag{5.5.2}$$

where α is the growth rate for the growing waves. From (5.5.1) and (5.5.2), the following energy conservation law is obtained:

$$\frac{dP}{dz} + \frac{dQ}{dz} = 0, \tag{5.5.3}$$

which means that the power increase for the electromagnetic wave mode is compensated for by the power decrease for the space-charge wave mode or the increase in the negative power carried by the latter. The increase in the negative power carried by the space-charge wave mode corresponds to the decrease in the kinetic power of the electron beam. Thus, through the process of active coupling between the electromagnetic wave mode and the space-charge wave mode, the Cherenkov laser amplifies the electromagnetic wave by converting the kinetic energy of the relativistic electron beam into the electromagnetic wave energy. As a result, the drift velocity of the electron beam gradually decreases as the electromagnetic wave gets energy from it and grows.

To calculate the change in the drift velocity of the electron beam as the electromagnetic wave propagates and grows along it, the beam-wave system is divided into many small segments in the direction of wave propagation (see Fig. 5.8). Then, we apply the linear analysis for each small segment, assuming a constant drift velocity of the electron beam in it, and investigate the increase in electromagnetic power in it. The drift velocity in the adjacent small segment is obtained by subtracting the increment in the electromagnetic power in the previous small segment from the power carried by the electron beam, which is defined as the kinetic energy of the electron beam carried per unit time across the plane perpendicular to it. By successively repeating this procedure, the growth of the electromagnetic wave and the decrease in the drift velocity of the electron beam are calculated.

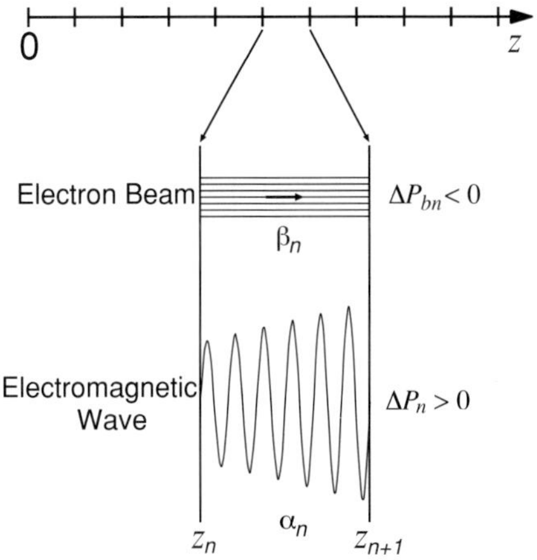

Fig. 5.8. Division of the beam-wave system into many small segments

If we assume that the drift velocity of the electron beam changes stepwise between adjacent segments with a constant value within each small segment, the power of the electromagnetic wave mode, $P_n(z)$ in the nth small segment is given by

$$P_n(z) = A_n \cosh^2 \alpha_n z, \tag{5.5.4}$$

where A_n and α_n are constants. The amplitude A_n must satisfy the condition that the electromagnetic power is continuous at the boundaries of adjacent segments, and $A_1 = P_0$.

On the other hand, the kinetic energy of one electron with the velocity equal to the average beam velocity V is expressed as

$$T = (\gamma - 1)m_0c^2, \quad \gamma = \frac{1}{\sqrt{1-\beta^2}}, \quad \beta = \frac{V}{c}. \tag{5.5.5}$$

Hence the power carried by the electron beam per unit width in the x direction is represented as

$$\begin{aligned} P_b &= (\gamma - 1)m_0c^2 NV\Delta_{eff} \\ &= m_0c^3 N\beta\Delta_{eff}\left(\frac{1}{\sqrt{1-\beta^2}} - 1\right), \end{aligned} \tag{5.5.6}$$

where N denotes the average number density of electrons, and Δ_{eff} the effective beam thickness, which is equal to the skin depth δ for thick beams and to the actual beam thickness for thin beams. The effective beam thickness is defined as the thickness of an equivalent beam across which the fields are uniform. In (5.5.6), we have the relation $N\beta = N_0\beta_0$ (constant), N_0 and β_0 representing the values of N and β at $z = 0$. This relation is derived from the continuity of the beam flow: the average number of electrons crossing the plane perpendicular to the electron beam per unit time is constant for all the segments.

Let the power increment for the electromagnetic wave mode in the nth segment be ΔP_n, and that for the space-charge wave mode ΔQ_n. Then, the following relation holds from (5.5.3):

$$\Delta P_n + \Delta Q_n = 0. \tag{5.5.7}$$

On the other hand, let the normalized drift velocity of the electron beam change from β_n to β_{n+1} due to the growth of the electromagnetic wave mode. Then, since the power increase in the space-charge wave mode carrying negative energy corresponds to the decrease in power carried by the electron beam, we get the following relation:

$$\Delta Q_n = \Delta P_{bn} = m_0 c^3 N \beta_n \Delta_{eff} \left[\frac{1}{\sqrt{1-\beta_{n+1}^2}} - \frac{1}{\sqrt{1-\beta_n^2}} \right], \tag{5.5.8}$$

where ΔP_{bn} denotes the variation in power P_b carried by the electron beam. From (5.5.7) and (5.5.8), we obtain the relation

$$\Delta P_n + m_0 c^3 N \beta_n \Delta_{eff} \left[\frac{1}{\sqrt{1-\beta_{n+1}^2}} - \frac{1}{\sqrt{1-\beta_n^2}} \right] = 0. \tag{5.5.9}$$

Solving (5.5.9) for the new normalized drift velocity β_{n+1}, we find

$$\beta_{n+1} = \sqrt{1 - \frac{1}{\gamma_{n+1}^2}}, \tag{5.5.10}$$

$$\gamma_{n+1} = -\frac{\Delta P_n}{m_0 c^3 N_0 \beta_0 \Delta_{eff}} + \frac{1}{\sqrt{1-\beta_n^2}}. \tag{5.5.11}$$

In this way, we can obtain a new normalized drift velocity β_{n+1} for each small segment, which is in turn used to calculate a new spatial growth rate α_{n+1}. Repeating the procedure outlined above successively, starting from $n = 0$, we can temporally follow the power transfer from the electron beam to the electromagnetic wave.

In Fig. 5.9 is illustrated a numerical example of the power growth for the electromagnetic wave versus the interaction distance. For comparison, the power growth for the case of a constant drift velocity is also shown in the figure. In Figs. 5.10 and 5.11 are given the variations of the drift velocity of the electron beam and the spatial growth rate versus the interaction distance, respectively. For the frequency ω, we have selected a value which initially makes the spatial growth rate

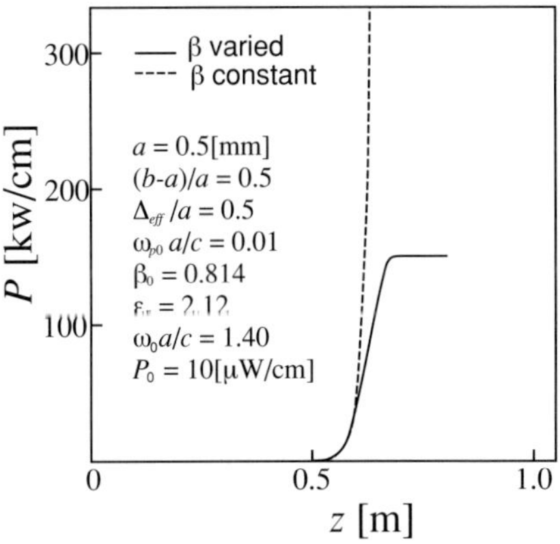

Fig. 5.9. Power growth for the electromagnetic wave

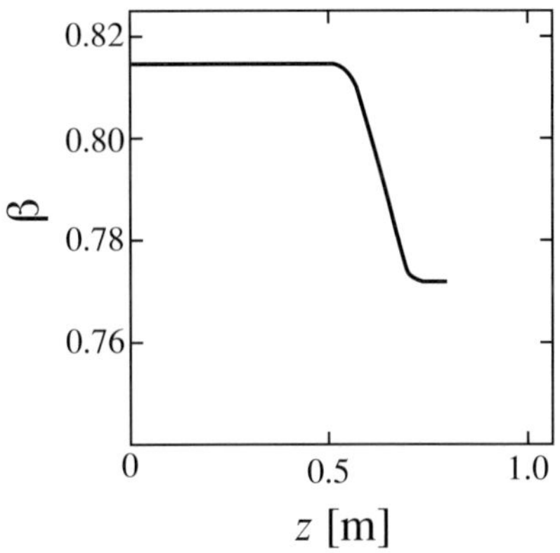

Fig. 5.10. Variation of the drift velocity

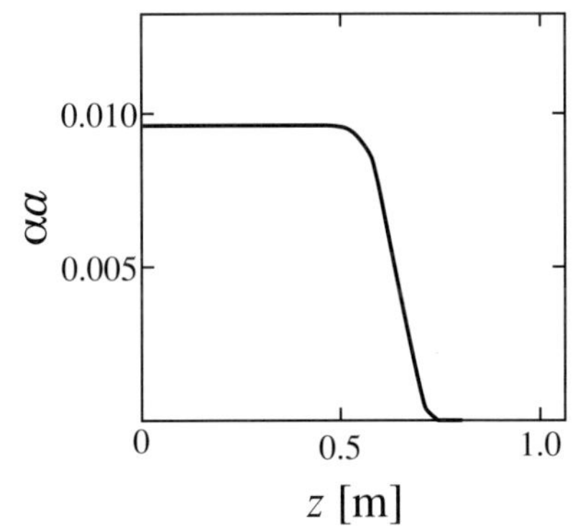

Fig. 5.11. Variation of the spatial growth rate

maximum. From these figures, we see that the power growth is suppressed by the decrease in the drift velocity of the electron beam. The decrease in the drift velocity causes a shift of the coupling point for the electromagnetic and space-charge wave modes, and reduces the spatial growth rate at the frequency selected initially, thus resulting in the smaller power growth compared with the case of a constant drift velocity. If we can reduce the phase velocity of the electromagnetic wave in accordance with the decrease in the drift velocity of the electron beam, the growth rate will remain maximum at the frequency selected initially, leading to a more efficient power growth for the electromagnetic wave. The phase velocity of the electromagnetic wave can be actually varied by changing the permittivity of the dielectric composing the waveguide.

5.6 Single-Particle Approach

As shown in Fig. 5.12, let us consider a two-dimensional model composed of a dielectric-coated conducting plane of infinite extent and a uniform linear distribution of electrons moving with constant velocity in the z diirection. Due to the 2-D geometry, each electron in Fig. 5.12 is regarded as a sample electron extracted from a uniform linear distribution of electrons of infinite extent in the x direction. For simplicity, we also assume that an infinite magnetostatic field is applied on the electron beam to restrict the motion of electrons only to the z direction. In the 2-D model of Fig. 5.12, we calculate the energy transfer from a linear distribution of electrons to the electromagnetic wave propagated along a dielectric-coated conducting plane waveguide, neglecting the space-charge effect and the thermal motion of electrons.

The energy relation, which describes the energy exchange between the electron and the electromagnetic field, is represented from (2.10.5) as

$$\frac{dK}{dt} = -ev_z E_z, \quad K = m_0 c^2 (\gamma - 1), \quad \gamma = \frac{1}{\sqrt{1-(v_z/c)^2}} \tag{5.6.1}$$

where K denotes the kinetic energy of the electron. In addition, from (2.10.6), the equation of motion for the electron under the condition specified above takes the form

$$\frac{dv_z}{dt} = -\frac{e}{\gamma^3 m_0} E_z . \tag{5.6.2}$$

Let the z component of electric field be expressed as

$$E_z = E_0 \cos(\omega t - k_z z + \phi), \tag{5.6.3}$$

where E_0 is a constant amplitude of the electric field, and ϕ is a phase denoting the position of a particular electron relative to the electric field. In addition, let the z

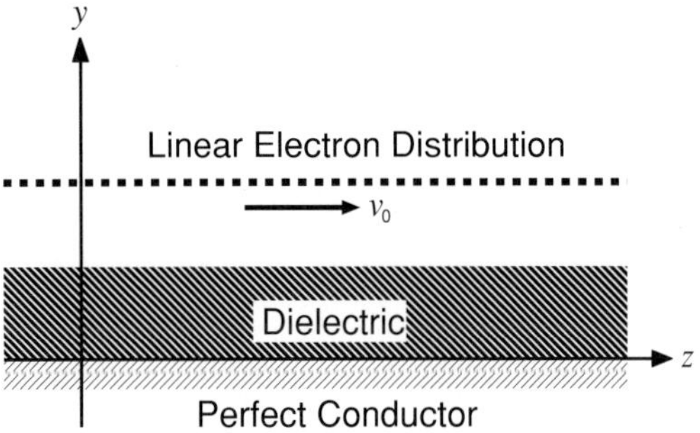

Fig. 5.12. A dielectric-coated conducting plane and a single-particle model of electron beam

coordinate of the electron at the time t be expressed as $z = v_0 t$ for the zeroth-order approximation. Then, the equation of motion for the electron (5.6.2) reduces, within the first-order approximation, to

$$\frac{dv_z}{dt} = -\frac{e}{\gamma_0^3 m_0} E_0 \cos(\Omega t + \phi), \quad \gamma_0 = \frac{1}{\sqrt{1-(v_0/c)^2}}, \tag{5.6.4}$$

$$\Omega = \omega - v_0 k_z, \tag{5.6.5}$$

Ω being the effective frequency measured at the position of the electron. Integrating the equation of motion for the electron (5.6.4) with respect to time, we get for the first-order value of the electron velocity,

$$v_z(t,\phi) = v_0 - \frac{eE_0}{\gamma_0^3 m_0 \Omega}\left[\sin(\Omega t + \phi) - \sin\phi\right]. \tag{5.6.6}$$

Integrating further (5.6.6) from 0 to t with respect to time, we obtain the first-order value of the electron position at time t,

$$z(t,\phi) = v_0 t + \frac{eE_0}{\gamma_0^3 m_0 \Omega^2}\left[\cos(\Omega t + \phi) - \cos\phi + \Omega t \sin\phi\right]. \tag{5.6.7}$$

The second-order value for the z component of electric field acting upon the electron at time t is obtained by inserting (5.6.7) in (5.6.3) as

$$E_z(t,\phi) = E_0 \cos(\Omega t + \phi)$$
$$+ \frac{e^2 k_z E_0^2}{\gamma_0^3 m_0 \Omega^2} \sin(\Omega t + \phi)\left[\cos(\Omega t + \phi) - \cos\phi + \Omega t \sin\phi\right]. \tag{5.6.8}$$

The energy equation for the electron, which is correct to the second-order with respect to small-signal fields, can be found by substituting (5.6.6) and (5.6.8) in (5.6.1) and leaving the resultant terms to the second-order in E_0:

$$\frac{dK}{dt} = -ev_0 E_0 \cos(\Omega t + \phi)$$
$$-\frac{ev_0 k_z E_0^2}{\gamma_0^3 m_0 \Omega^2} \sin(\Omega t + \phi)\left[\cos(\Omega t + \phi) - \cos\phi + \Omega t \sin\phi\right]$$
$$+\frac{e^2 E_0^2}{\gamma_0^3 m_0 \Omega} \cos(\Omega t + \phi)\left[\sin(\Omega t + \phi) - \sin\phi\right]. \tag{5.6.9}$$

Averaging (5.6.9) over electrons with phase ϕ relative to the electric field, which is uniformly distributed from 0 to 2π, we get

$$\left\langle \frac{dK}{dt} \right\rangle_\phi \equiv \frac{1}{2\pi}\int_0^{2\pi} \frac{dK}{dt} d\phi$$
$$= \frac{\omega e^2 E_0^2}{2\gamma_0^3 m_0 \Omega^2}\left[\sin \Omega t - \Omega t \cos \Omega t\right], \tag{5.6.10}$$

where the condition $|\Omega| << \omega$ is assumed. Equation (5.6.10) represents the average amount of energy gained by the electron if it is positive, and that lost by the electron if it is negative. Hence we have for the average variation of energy carried by the electron when it has traveled the distance L,

$$\langle \Delta K \rangle_\phi = \int_0^{L/v_0} \left\langle \frac{dK}{dt} \right\rangle_\phi dt = \frac{\omega e^2 \tau^3 E_0^2}{2\gamma_0^3 m_0} g(\Omega\tau), \tag{5.6.11}$$

with

$$g(\Omega\tau) = \frac{2(1-\cos \Omega\tau) - \Omega\tau \sin \Omega\tau}{(\Omega\tau)^3}, \tag{5.6.12}$$

where $\tau = L/v_0$ and $g(\Omega\tau)$ denotes the gain function, which is a well-known function in the theory of free-electron lasers. The function g is an antisymmetric function, and its absolute value takes the maximum value 0.135 at $\Omega\tau = \pm 2.6$ (see Fig. 5.13). As seen from (5.6.11), the average increment of the kinetic energy of electrons becomes negative for $g(\Omega\tau) < 0$, while it becomes positive for $g(\Omega\tau) > 0$. In other words, in the region where $g(\Omega\tau) < 0$, the ensemble of electrons loses energy to the electromagnetic wave interacting with it, as a result of which the latter is amplified. On the other hand, in the region where $g(\Omega\tau) > 0$, the ensemble of electrons gains energy from the electromagnetic wave, which, in turn, gets damped. The main gain for the electromagnetic wave occurs for $\Omega\tau < 0$, which means

$$\omega - v_0 k_z < 0, \tag{5.6.13}$$

or

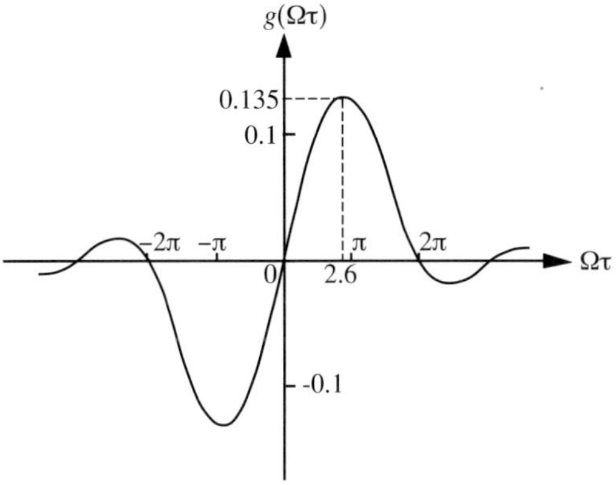

Fig. 5.13. Gain function $g(\Omega\tau)$

$$v_p < v_0, \quad v_p = \omega / k_z, \tag{5.6.14}$$

where v_p denotes the phase velocity of the electromagnetic wave. Equation (5.6.14) is just the condition for the Cherenkov radiation to occur, which has been clarified in the foregoing discussion (see Sect. 3.8). The amount of energy, which the electromagnetic wave gains from the electron beam, becomes maximum at $\Omega\tau = -2.6$.

With the aid of (5.6.11), let us try to find a specific expression for the power gain of the electromagnetic wave propagated along a dielectric-coated conducting plane waveguide shown in Fig. 5.12. Referring to the derivation in Sect. 5.2, we can express the field components of the electromagnetic wave mode in the dielectric-conductor waveguide as

$$\tilde{E}_z = A \sin p_y y, \tag{5.6.15}$$

$$\tilde{E}_y = -j \frac{k_z}{p_y} A \cos p_y y, \tag{5.6.16}$$

$$\tilde{H}_x = j \frac{\omega \varepsilon}{p_y} A \cos p_y y, \tag{5.6.17}$$

for $0 < y < a$, and

$$\tilde{E}_z = C e^{-h_y y}, \tag{5.6.18}$$

$$\tilde{E}_y = -j \frac{k_z}{h_y} C e^{-h_y y}, \tag{5.6.19}$$

$$\tilde{H}_x = j \frac{\omega \varepsilon_0}{h_y} C e^{-h_y y}, \tag{5.6.20}$$

for $y > a$, where ε denotes the permittivity of the dielectric, and the constant amplitudes A and C are related by

$$A = C \frac{e^{-h_y a}}{\sin p_y a}, \tag{5.6.21}$$

which can be found from the boundary condition at $y = a$. The wave numbers p_y and h_y are defined in (5.3.2) and (5.3.7), and the wave number k_z is obtained from (5.4.8). The electromagnetic power propagated along the dielectric-conductor waveguide per unit length in the x direction is calculated by integrating the z component of the Poyinting vector over a plane perpendicular to the z-axis, being expressed as

$$P = \mathrm{Re}\left[-\frac{1}{2}\int_0^\infty \tilde{E}_y \tilde{H}_x^* dy\right] = C^2 \frac{\omega k_z}{4p_y^2}\left[\frac{\varepsilon - \varepsilon_0}{h_y^3}\left(\frac{\omega}{c}\right)^2 + \frac{\varepsilon a}{\sin^2 p_y a}\right] e^{-2h_y a}. \tag{5.6.22}$$

On the other hand, the power increment for the electromagnetic wave corresponding to the average variation of the electron energy given by (5.6.11) is represented as

$$\langle \Delta P \rangle = -\frac{I}{e}\langle \Delta K \rangle_\phi , \tag{5.6.23}$$

where I denotes the magnitude of the current carried by the flow of electrons per unit width in the x direction. In addition, from (5.6.18) and (5.6.22), the magnitude of the longitudinal electric field E_0 at $y = b$ is expressed in terms of the power carried by the electromagnetic wave as

$$E_0^2 = C^2 e^{-2h_y b} = \frac{P}{F(\omega, k_z)} e^{-2h_y (b-a)}, \tag{5.6.24}$$

where

$$F(\omega, k_z) = \frac{\omega k_z}{4p_y^2}\left[\frac{\varepsilon - \varepsilon_0}{h_y^3}\left(\frac{\omega}{c}\right)^2 + \frac{\varepsilon a}{\sin^2 p_y a}\right]. \tag{5.6.25}$$

The power gain G for the electromagnetic wave propagated along the dielectric-conductor waveguide can be obtained from (5.6.23), (5.6.11), and (5.6.24) as follows:

$$G = \frac{\langle \Delta P \rangle}{P} = -\frac{\omega e I \tau^3}{2\gamma_0^3 m_0 F(\omega, k_z)} e^{-2h_y (b-a)} g(\Omega\tau), \tag{5.6.26}$$

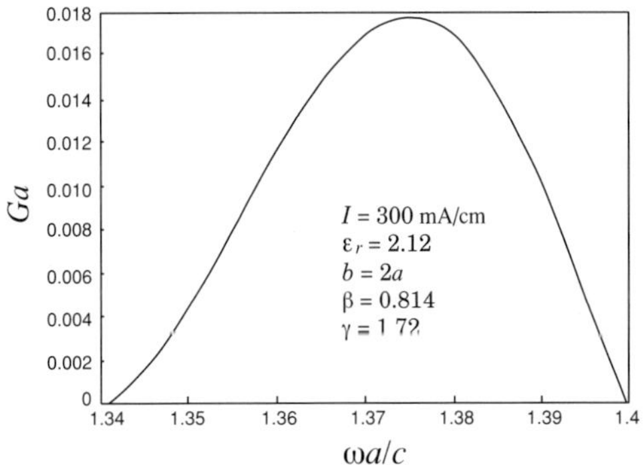

Fig. 5.14. Power gain for the electromagnetic wave mode

where we must insert the value of k_z which satisfies the dispersion relation (5.4.8) for a given value of ω. In Fig. 5.14, the frequency dependence of the power gain for the TM_0 mode is illustrated.

5.7 Trapping of Electrons in Electric Field

In the foregoing section, we have discussed the interaction of a relativistic electron beam and an electromagnetic wave on the basis of single-particle approach, together with quasi-linear approximations, finding that a strong beam-wave interaction occurs for the drift velocity of the electron beam nearly equal to the phase velocity of the electromagnetic wave. In this section, let us consider the beam-wave interaction in a more general manner which allows a nonlinear treatment of the problem. Before entering a detailed discussion, we briefly discuss how a distribution of electrons, which is uniform along the z-axis at the initial state, changes as the beam-wave interaction proceeds. Let us first assume that a uniform distribution of electrons is drifting with the velocity equal to the phase velocity of the electromagnetic wave at the initial state (see Fig. 5. 15(a)). Then, as time passes, electrons in the accelerating phases of the electric field are accelerated and pushed forward, while those in the decelerating phases of the electric field are decelerated and lag behind. As a result, dense and sparse parts in the distribution of electrons appear alternately at the same period as that of the electromagnetic wave, as shown in Fig. 5. 15(b). In other words, electrons are density-modulated or bunched along the electron beam. For the case where the electron beam and the electromagnetic wave are traveling with the same velocity, the amount of accelerated electrons turns out to be equal to that of decelerated electrons. Hence no net transfer of energy occurs between the electron beam and the electromagnetic wave. On the other hand, if a distribution of electrons is drifting with a velocity slightly greater than the phase velocity of the electromagnetic wave, the dense parts in the distribution of electrons shift toward the decelerating phases of the electric field, while the

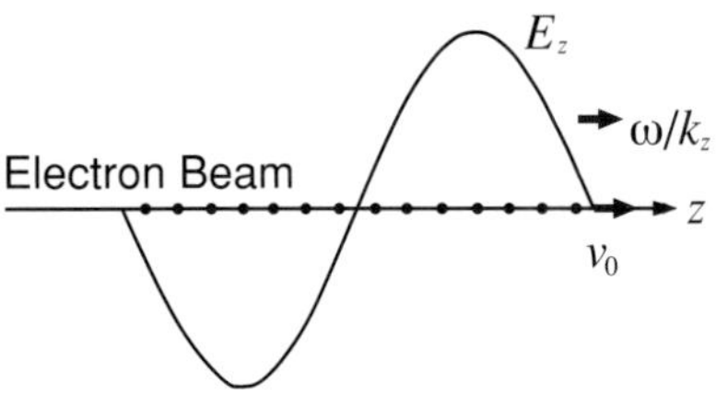

(*a*) $t = 0$

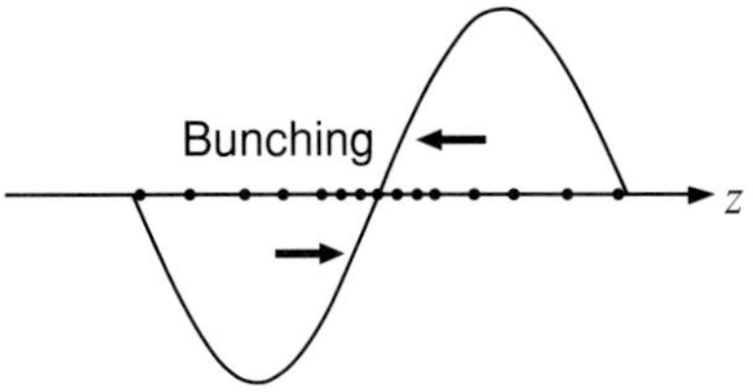

(*b*) $t > 0$ $(v_0 = \frac{\omega}{k_z})$

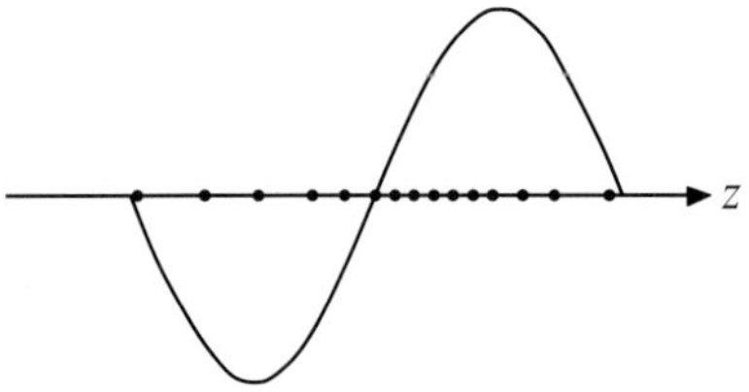

(*c*) $t > 0$ $(v_0 > \frac{\omega}{k_z})$

Fig. 5.15. Bunching of electrons

sparse parts shift toward the accelerating phases, as seen from Fig. 5. 15(*c*). Then more electrons are decelerated, and thus the net energy is transferred from the electron beam to the electromagnetic wave. Similarly, if a distribution of electrons is drifting with a velocity slightly less than the phase velocity of the electromagnetic wave, net energy is transferred from the electromagnetic wave to the electron beam.

We are now ready to discuss bunching and trapping of electrons in the electric field in a more general manner. The basic equations for the following discussion are the energy relation for the electron (5.6.1) and the relativistic equation of motion for the electron (5.6.2). First, we rewrite the left-hand side of (5.6.1) as

$$\frac{dK}{dt} = \frac{dK}{dz}\frac{dz}{dt} = v_z \frac{dK}{dz}. \tag{5.7.1}$$

From (5.6.1) and (5.7.1), we get the relation

$$\frac{dK}{dz} = -eE_z = -eE_0 \cos(k_z z - \omega t - \phi). \tag{5.7.2}$$

Similarly, we have from (5.6.2),

$$\frac{dv_z}{dz} = -\frac{e}{\gamma^3 v_z m_0} E_z = -\frac{e}{\gamma^3 v_z m_0} E_0 \cos(k_z z - \omega t - \phi). \tag{5.7.3}$$

In order to describe the behavior of a particular electron in the electric field, we define a new phase ψ as

$$\psi = k_z z - \omega t - \phi + \frac{\pi}{2}, \tag{5.7.4}$$

from which we find the following relation:

$$\frac{d\psi}{dz} = k_z - \frac{\omega}{v_z}. \tag{5.7.5}$$

Introducing the resonant velocity for the electron v_r, we rewrite v_z as

$$v_z = v_r + \Delta v_z, \quad v_r = v_p = \frac{\omega}{k_z}, \tag{5.7.6}$$

where v_p is the phase velocity of the electromagnetic wave, and we assume that $|\Delta v_z| << v_r$. Inserting (5.7.6) in (5.7.5), Eq. (5.7.5) reduces to

$$\frac{d\psi}{dz} = \frac{\omega}{v_r}\frac{\Delta v_z}{v_r} = k_z \frac{\Delta v_z}{v_r}. \tag{5.7.7}$$

In view of the relation $\Delta\gamma = \gamma^3 v_z / c^2 \Delta v_z$, Eq. (5.7.7) can also be expressed as

$$\frac{d\psi}{dz} = k_z \frac{c^2}{\gamma_r^2 v_r^2} \frac{\Delta\gamma}{\gamma_r}, \quad \gamma_r = \frac{1}{\sqrt{1 - \left(\frac{v_r}{c}\right)^2}}. \tag{5.7.8}$$

Similarly, from (5.7.5), (5.7.3), (5.7.4), and (5.7.6), we get the following relation:

$$\frac{d^2\psi}{dz^2} = -k_\psi^2 \sin\psi, \tag{5.7.9}$$

where

$$k_\psi^2 = \frac{\omega e E_0}{\gamma_r^3 v_r^3 m_0}. \tag{5.7.10}$$

Equation (5.7.9) is referred to as the pendulum equation, because it is of the same form as the equation for describing the motion of a pendulum swinging in a gravitational field [5.13].

With the aid of (5.7.7) or (5.7.8), and (5.7.9), we can describe the behavior of a particular electron with a particular phase relative to the electric field in a general manner. For small ψ, the pendulum equation (5.7.9) can be easily solved, proving to represent a simple harmonic motion. To solve (5.7.9) for a general case, multiply both sides of it by $2(d\psi/dz)$ and integrate with respect to z, and we get the following equation:

$$\left(\frac{d\psi}{dz}\right)^2 = 2k_\psi^2\left(\cos\psi + C\right), \tag{5.7.11}$$

or

$$\frac{d\psi}{dz} = \pm\sqrt{2}k_\psi\sqrt{\cos\psi + C}, \tag{5.7.12}$$

where C is an integration constant, which is determined by the boundary conditions. The variation in the velocity and kinetic energy of the electron can be evaluated from (5.7.12), with the aid of (5.7.7) or (5.7.8). If we adopt the boundary conditions that $d\psi/dz = 0$ at $\psi = \pm\pi$, we obtain $C = 1$. Then, we have

$$\frac{d\psi}{dz} = \pm\sqrt{2}k_\psi\sqrt{\cos\psi + 1}. \tag{5.7.13}$$

The curve given by (5.7.13) is referred to as the separatrix, which separates the region of closed orbits of electrons and the region of their open orbits. The region for $C > 1$ corresponds to open orbits, while the region for $C < 1$ is for closed orbits. In Fig. 5.16, we illustrate some examples of electron orbits on the pendulum phase diagram, in which the coordinate $d\psi / dz$ is plotted versus the coordinate ψ for the phase angles between $-\pi$ and 3π. As is evident from (5.7.7) or (5.7.8), the coordinate $d\psi / dz$ is proportional to the variation in the velocity or kinetic energy of the electron. On the other hand, the coordinate ψ represents the position of the electron at a particular time relative to the distribution of the electric field. The plus sign in the right-hand side of (5.7.12) corresponds to the upper half of the phase diagram while the minus sign to the lower half of the diagram. From (5.7.13), we see that the maximum width of the separatrix in the vertical direction is $4k_\psi$, which is proportional to the square root of the magnitude of longitudinal electric field. Therefore, the maximum width of closed orbits is less than this value. For refernce, we also show in the figure the distribution of the longitudinal electric field. As seen from the figure, the left half of the diagram ($-\pi$, 0) is in the accelerating phase for the electron while the right half (0, π) is in the decelerating phase.

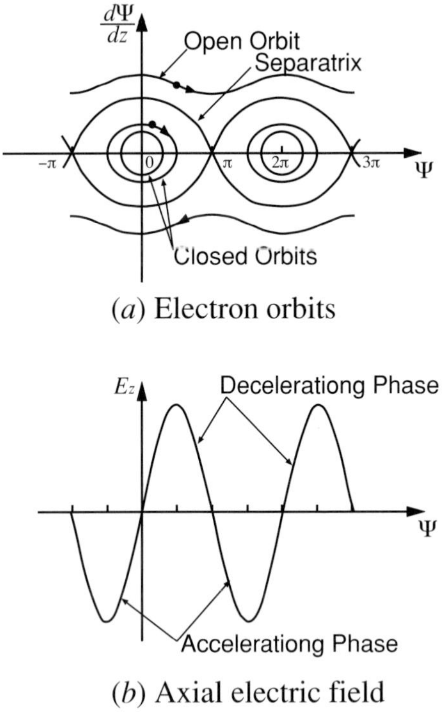

Fig. 5.16. Electron orbits on the phase diagram

Let us further proceed with (5.7.12) for the closed orbits. For this purpose, we rewrite (5.7.12) with the aid of the boundary conditions. If we use the boundary conditions that $d\psi / dz = 0$ at $\psi = \pm\psi_0 (0 < \psi_0 < \pi)$, we have

$$\frac{d\psi}{dz} = \pm\sqrt{2}k_\psi \sqrt{\cos\psi - \cos\psi_0}\,, \tag{5.7.14}$$

which can be rewritten as

$$\frac{d\psi}{dz} = \pm 2k_\psi \sqrt{\sin^2\frac{\psi_0}{2} - \sin^2\frac{\psi}{2}}\,. \tag{5.7.15}$$

To modify (5.7.15), we change variables from ψ to u according to

$$\sin\frac{\psi}{2} = \eta \sin u\,, \quad \eta = \sin\frac{\psi_0}{2}\,, \tag{5.7.16}$$

where $0 < \eta < 1$. Then, we have

$$\frac{du}{dz} = \pm k_\psi \sqrt{1 - \eta^2 \sin^2 u}\,. \tag{5.7.17}$$

Let the spatial period for an orbital motion of the electron be denoted by L_ψ. Then, from (5.7.17), we get the relation

$$\int_0^{L_\psi} k_\psi dz = 2\int_{-\pi/2}^{\pi/2} \frac{du}{\sqrt{1-\eta^2 \sin^2 u}} = 4\int_0^{\pi/2} \frac{du}{\sqrt{1-\eta^2 \sin^2 u}}, \tag{5.7.18}$$

where the right-hand side denotes an elliptic integral of the first kind. For small η, we can calculate the elliptic integral by expanding the integrand in a power series with respect to η^2, finding the following result:

$$\int_0^{L_\psi} k_\psi dz = 2\pi\left(1 + \frac{1}{4}\eta^2 + \frac{9}{64}\eta^4 + \cdots\cdots\right), \tag{5.7.19}$$

from which we get

$$L_\psi = \frac{2\pi}{k_\psi}\left(1 + \frac{1}{4}\eta^2 + \frac{9}{64}\eta^4 + \cdots\cdots\right). \tag{5.7.20}$$

For a special case of sufficiently small values of η, which corresponds to a simple harmonic motion, we obtain

$$L_\psi = \frac{2\pi}{k_\psi} = 2\pi\left(\frac{\gamma_r^3 v_r^3 m_0}{\omega e E_0}\right)^{1/2}. \tag{5.7.21}$$

On the basis of the above preparations, we discuss the behavior of electrons in the wave field. For a typical case, let us consider, at the initial state, a longitudinal electric field with a sufficiently small amplitude and a uniform distribution of electrons drifting with a velocity slightly greater than the phase velocity of the electromagnetic wave. As stated earlier in this section, electron bunches are formed in the decelerating phases of the electric field as the beam-wave interaction proceeds, and the wave amplitude gradually grows as energy is transferred from the electron beam to the electromagnetic field. While the wave amplitude is sufficiently small, electrons travel on open orbits as given by (5.7.12) with $C > 1$. However, when it becomes sufficiently large as the beam-wave interaction continues further, bunched electrons get trapped in the electric field. Thereafter, bunched electrons move around along closed orbits as given by (5.7.12) with $C < 1$, in the phase space of electric field.

To have a qualitative idea as to the beam-wave interaction, we illustrate in Fig. 5.17 the behavior of electrons in the phase diagram. At the initial state, electrons with the same velocity are uniformly distributed in the phase of electric field between $-\pi$ and π. Then, forces exerted on electrons are different from electron to electron, and each electron moves on its own orbit. As the interaction between the ensemble of electrons and the electric field proceeds, electrons initially arranged in

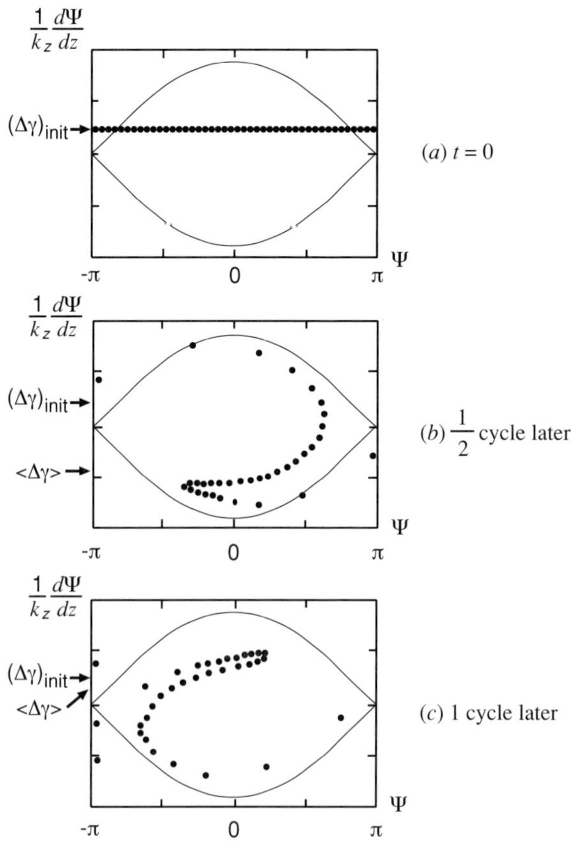

Fig. 5.17. Behavior of electrons in the phase space

a straight line get dispersed in the phase space. About a half cycle of orbital motion later, as seen from Fig. 5.17, the kinetic energy averaged over the ensemble of electrons becomes minimum. About a full cycle of orbital motion later, in turn, it becomes maximum, returning to almost the same level as at the initial state. After that, the same cyclic motion continues. Hence we find that the maximum energy can be extracted from the electron beam after about a half cyclic motion of electrons in the phase space. We should also note that the velocity and energy distributions for the ensemble of electrons are spread as a necessary effect of the wave-electron interaction, even when their distributions due to their thermal motion are negligible.

6. Single-Particle Theory of the Free-Electron Laser

6.1 Introduction

The seeds of the research on free-electron lasers can be found in the research of microwave electron tubes in the 1950's. In 1951, Motz first analyzed the frequency spectrum of electromagnetic waves emitted from an electron traveling in an array of permanent magnets called an undulator [6.4]. Immediately after that, Phillips succeeded in generating high-power millimeter waves, using an undulator [6.5]. Since then, we had to wait for a while before the advent of the first free-electron laser at Stanford in the middle 1970's [6.6-6.8]. The free-electron laser operates on a mechanism different from that for the conventional laser. The ordinary laser utilizes coherent radiation emitted when electrons in a lasing material medium drop from higher energy levels to lower ones. On the other hand, the free-electron laser produces coherent radiation by directly converting the kinetic energy of electrons, which are moving with relativistic velocities in a periodic static magnetic field, to electromagnetic wave energy. As shown in Fig. 6.1, a typical free-electron laser is composed of three main components, namely, an electron accelerator, an array of permanent magnets (called a wiggler or an undulator), and a resonator consisting of a pair of reflecting mirrors. Amplification of electromagnetic waves in the free-electron laser occurs as a result of the interaction between an ensemble of relativistically moving electrons and an electromagnetic wave propagated with the former in the same direction, under the influence of a transverse periodic magneto-static field produced by an array of permanent magnets.

First, we briefly summarize the properties of radiation emitted from an electron undulating in an array of permanent magnets. An electron traveling along the center axis of an undulator undergoes an undulating motion with the same spatial period as that of the magnetic field produced by the undulator, in the plane perpendicular to the magnetic field. Then, every time the orbit of an electron is bent by the magnetic field, it emits synchrotron radiation. An observer in the far distance along the direction of motion of the electron will see a superposition of electromagnetic wave pulses emitted at each corner of the electron orbit. As the number of electromagnetic wave pulses or the number of undulation of the electron increases, the frequency spectrum for a superposition of electromagnetic wave pulses will approach that for a monochromatic wave, consisting of fundamental and higher har-

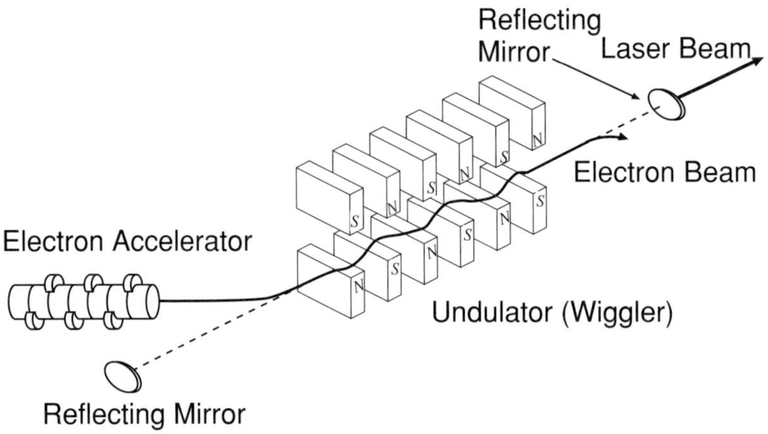

Fig. 6.1. Structural view of the free-electron laser

monic waves. The above discussion applies only to the radiation emitted by an individual electron in the ensemble of electrons, while the radiation produced by the whole ensemble of electrons becomes a superposition of electromagnetic waves radiated by individual electrons composing it. In general, velocities of individual electrons are randomly distributed around a particular average value. Hence a superposition of electromagnetic waves emitted from these individual electrons is temporally incoherent, while it is spatially coherent because synchrotron radiation is sharply directed in the direction of motion of electrons. From the above consideration, synchrotron radiation produced in an undulator is found to correspond to spontaneous emission in the conventional laser theory.

Next, let us consider the interaction of a plane electromagnetic wave with the ensemble of relativistically moving electrons, under the influence of a transverse periodic magnetostatic field. When an electromagnetic wave is propagated in a transverse periodic magnetic field, they are coupled to generate a beat wave. The beat wave has the same frequency as the electromagnetic wave, but a wave number which is equal to the sum of the wave numbers for the electromagnetic wave and the periodic static magnetic field. This means that the beat wave has a wave number larger than the electromagnetic wave. In other words, the wavelength of the beat wave becomes shorter than that of the electromagnetic wave. Thus the beat wave can propagate with a velocity slower than that of light. Then, how can we get a longitudinal force to cause electrons to be decelerated and to release their kinetic energy to the electromagnetic wave, from the purely transverse electromagnetic and static magnetic fields? When an electron travels in a magnetic field, it is subject to the Lorentz force which is perpendicular to both the direction of motion and the magnetic field. Hence a transverse force is exerted on an electron injected in a transverse magnetostatic field, giving a transverse velocity to it. Then, this transverse velocity given to the electron and the transverse magnetic field of the electromagnetic wave make a longitudinal force acting on it, which corresponds to the force due to the beat wave described above.

Electromagnetic waves produced by the process of spontaneous emission due to the ensemble of electrons relativistically moving in un undulator cover a wide range of frequency spectrum. Of these wide-range wave components, those which produce beat waves with a propagation velocity nearly equal to the average velocity of electrons strongly interact with the ensemble of electrons. As a result of this resonant interaction between the electromagnetic wave and electrons, the ensemble of electrons is density-modulated by the longitudinal force due to the beat wave. Through this process, electrons are bunched in the longitudinal direction. If the beat wave travels in the longitudinal direction with the same velocity as the average longitudinal velocity of the ensemble of electrons, the numbers of electrons accelerated and decelerated by bunching become equal, which does not lead to net transfer of energy between the electromagnetic wave and the ensemble of electrons. This situation occurs for the center frequency of the spectrum for spontaneous emission by electrons described in the foregoing discussion. On the other hand, if the ensemble of electrons travels slightly faster than the beat wave, more electrons are decelerated, causing amplification of the electromagnetic wave. This case corresponds to stimulated emission. Furthermore, for the case where the ensemble of electrons travels slightly slower than the beat wave, more electrons are accelerated with the electromagnetic wave damped, which corresponds to stimulated absorption. As will be clarified later, the stimulated emission occurs for frequencies slightly lower than the center frequency of the spectrum for spontaneous emission, while the stimulated absorption occurs for frequencies slightly higher than the center frequency. For the ensemble of electrons traveling on the average faster than the beat wave, individual electrons randomly distributed in the phase of the beat wave at the initial state gradually get bunched in the decelerating phase of the beat wave, and finally behave like a macroparticle as a whole. As a result, we can get a coherent electromagnetic wave by the stimulated emission due to the bunched electrons. In the following sections, we discuss in more detail each process leading to the stimulated emission in the free-electron laser.

6.2 Synchrotron Radiation from an Array of Permanent Magnets

6.2.1 Condition for Constructive Interference

For simplicity, let us consider a 2-dimensional array of permanent magnets as shown in Fig. 6.2. On the center plane of the 2-D undulator, the transverse static magnetic field varies with the spatial period λ_0 in the longitudinal z direction. The transverse y component of the magnetic-flux density for the periodic static magnetic field is expressed as

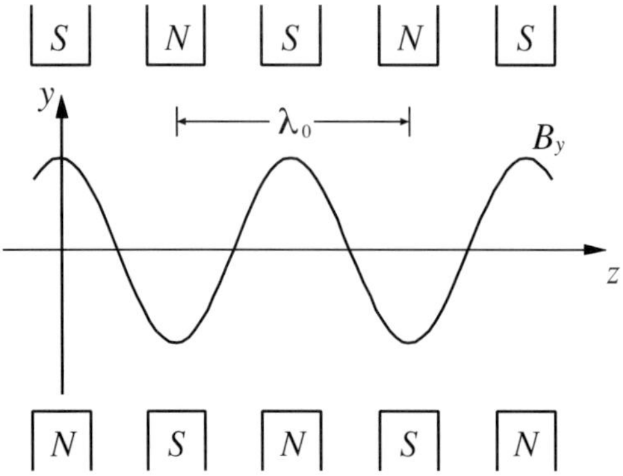

Fig. 6.2. 2-dimensional array of permanent magnets

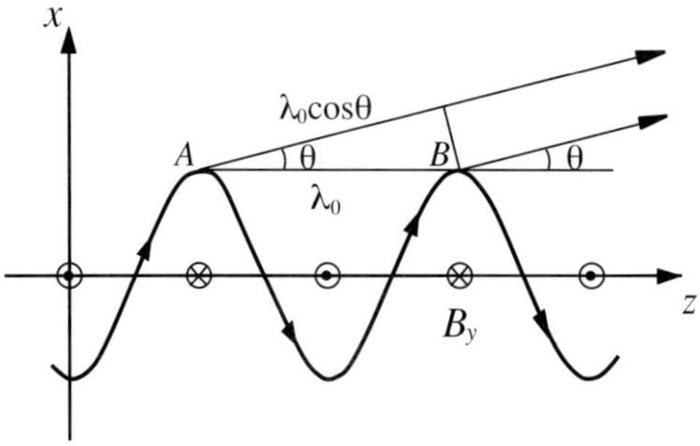

Fig. 6.3. Electron orbit in the center plane of an undulator

$$B_y = B_0 \cos k_0 z\,, \quad k_0 = \frac{2\pi}{\lambda_0}. \tag{6.2.1}$$

In the first-order approximation, an electron traveling along the z-axis with a relativistic velocity undergoes an undulating motion around the z-axis on the center plane perpendicular to, and with the same spatial period as, the transverse static magnetic field, as illustrated in Fig. 6.3. Then, every time the orbit of the electron is bent by the magnetic field, it emits synchrotron radiation for which the angular distribution of the frequency spectrum is given by (3.7.15).

We first discuss the condition under which an electromagnetic wave pulse emitted at each corner of the electron orbit interferes with one another constructively. For this purpose, let us compare the transit distances of two pulses emitted from two adjacent corners of the electron orbit, A and B in Fig. 6.3. Let the electron emit an optical pulse at the point A at the time 0, and a second pulse at B after the time $\tau_0 = \lambda_0 / \bar{v}_z$, where $\bar{v}_z$ denotes the average longitudinal velocity of the electron. Then, if the difference in the transit distances for two optical pulses is an integral multiple of the optical wavelength, two optical pulses interfere with each other constructively. For the special case where the difference in transit distance is equal to one optical wavelength, the following relation holds:

$$\lambda_s = c\frac{\lambda_0}{\bar{v}_z} - \lambda_0 \cos\theta, \tag{6.2.2}$$

where λ_s denotes the wavelength of the optical wave, and θ is an angle between the direction of pulse propagation and the z-axis.

In order to find the value of $\bar{v}_z$, we first calculate the undulating velocity of the electron, v_x. From the relativistic equation of motion for the electron, (2.10.6) and (6.2.1), the undulating velocity v_x is found to satisfy

$$\frac{dv_x}{dt} = \frac{eB_0}{\gamma m_0} v_z \cos k_0 z, \tag{6.2.3}$$

with

$$\gamma = \frac{1}{\sqrt{1-\beta^2}}, \quad \beta = \frac{v}{c}, \tag{6.2.4}$$

where v denotes the electron velocity along its orbit. With the aid of the relation $v_z = dz/dt$, we can readily solve (6.2.3) to get

$$v_x = \frac{eB_0}{\gamma m_0 k_0} \sin k_0 z. \tag{6.2.5}$$

Using (6.2.5) and (6.2.4), we can express v_z as

$$\begin{aligned} v_z &= \sqrt{v^2 - v_x^2} \\ &= c\sqrt{1 - \frac{1}{\gamma^2}\left(1 + a_u^2 \sin^2 k_0 z\right)} \\ &\approx c\left[1 - \frac{1}{2\gamma^2}\left(1 + a_u^2 \sin^2 k_0 z\right)\right], \end{aligned} \tag{6.2.6}$$

where

$$a_u = \frac{eB_0}{m_0 k_0 c}. \tag{6.2.7}$$

In the third line of (6.2.6), we have assumed $\gamma >>1$. Averaging (6.2.6) with respect to z over one spatial period of the undulator, we get

$$\bar{v}_z = c\left[1 - \frac{1}{2\gamma^2}\left(1 + \frac{1}{2}a_u^2\right)\right]. \tag{6.2.8}$$

Let us consider the case of $\gamma >>1$ and $\theta <<1$, which we exclusively treat in the subsequent discussion. Then, from (6.2.2) and (6.2.8), we obtain for the final result for the relation between λ_s and λ_0,

$$\lambda_s = \frac{\lambda_0}{2\gamma^2}\left[1 + \frac{1}{2}a_u^2 + (\gamma\theta)^2\right]. \tag{6.2.9}$$

The frequency relation corresponding to (6.2.9) is given by

$$\omega_s = \frac{2\gamma^2\omega_0}{1 + (1/2)a_u^2 + (\gamma\theta)^2}, \quad \omega_0 = \bar{v}_z k_0, \tag{6.2.10}$$

where ω_s is the angular frequency of the optical wave which satisfies the condition for constructive interference, and ω_0 denotes the angular frequency in the laboratory frame with which the electron feels the transverse periodic static magnetic field.

6.2.2 Frequency Spectrum

Let us discuss the frequency spectrum for synchrotron radiation observed from the far distance on the z-axis. Setting $\theta = 0$ in (3.7.15), we have the frequency spectrum for synchrotron radiation emitted from a single curved orbit of the electron per unit solid angle in the longitudinal z direction

$$\frac{d^2F(\omega,\theta)}{d\Omega d\omega} = \frac{3e^2}{4\pi^3}\sqrt{\frac{\mu_0}{\varepsilon_0}}\gamma^2\left(\frac{\omega}{\omega_{cr}}\right)^2 K_{2/3}^2\left(\frac{\omega}{\omega_{cr}}\right), \tag{6.2.11}$$

where

$$\omega_{cr} = 3\gamma^3\left(\frac{c}{\rho}\right), \tag{6.2.12}$$

ρ being the instantaneous radius of curvature for the electron orbit. At each corner of the electron orbit, in particular, ω_{cr} takes the value

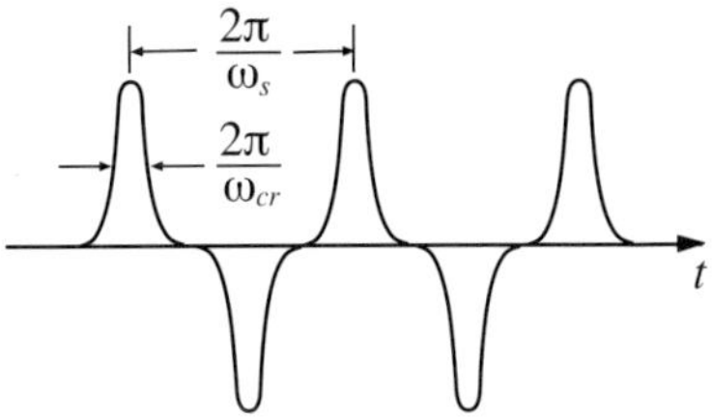

Fig. 6.4. Pulse train observed for $\theta = 0$

$$\omega_{cr} \approx 3\gamma^3 \omega_c = 3\gamma^2 \frac{eB_0}{m_0} \tag{6.2.13}$$

where ω_c denotes the cyclotron frequency for the electron. For $\theta = 0$, it should be noted that only waves polarized in the plane of the electron orbit are emitted. Now, let us calculate the frequency spectrum for synchrotron radiation emitted from an undulator composed of $2N$ pairs of permanent magnet poles polarized in the opposite directions alternately. Referring to (3.7.12), we see that we get opposite polarities alternately for the waves polarized parallel to the orbital plane for the electron as the electron passes through a series of magnet poles. For this case, we have a pulse train as shown in Fig. 6.4, which is composed of the sum of $2N$ optical pulses with alternate polarities emitted every time period of π/ω_s. For simplicity, we assume that individual pulses are clearly separated. If the width of each pulse is much less than the period of pulse repetition, we can get a train of clearly separated pulses. This condition is quantitatively expressed as $\omega_{cr} >> \omega_s$, which is found to be equivalent to $a_u >> 1$. The frequency spectrum for synchrotron radiation emitted from an electron passing through $2N$ pairs of permanent magnets per unit solid angle in the z direction can be calculated by superposing each pulse emitted at each corner of the electron orbit, taking the polarity and time delay of each pulse into account. Hence the frequency spectrum for the synchrotron radiation from a single electron undulating through the array of permanent magnets can be expressed as

$$\frac{d^2F(\omega,\theta)}{d\Omega d\omega} = \frac{3e^2}{4\pi^3}\sqrt{\frac{\mu_0}{\varepsilon_0}}\gamma^2\left(\frac{\omega}{\omega_{cr}}\right)^2 K_{2/3}^2\left(\frac{\omega}{\omega_{cr}}\right)\left|\sum_{n=0}^{2N-1}(-1)^n e^{-j\omega\frac{n\pi}{\omega_s}}\right|^2 . \tag{6.2.14}$$

The summation in (6.2.14), which is denoted by S, can be readily calculated as

$$|S|^2 = \frac{1-\cos\left[2N\pi\left(\frac{\omega}{\omega_s}\right)\right]}{1+\cos\left[\pi\left(\frac{\omega}{\omega_s}\right)\right]} = \frac{\sin^2\left[N\pi\left(\frac{\omega}{\omega_s}-2m-1\right)\right]}{\sin^2\left[\frac{\pi}{2}\left(\frac{\omega}{\omega_s}-2m-1\right)\right]},$$

$$(m = 0, 1, 2, \cdots). \tag{6.2.15}$$

The denominator of (6.2.15) varies much more slowly than the numerator for large values of N. Then, it can be approximated around $\omega / \omega_s = 2m + 1$ as

$$|S|^2 = 4N^2 \frac{\sin^2\left[N\pi\left(\dfrac{\omega}{\omega_s} - 2m - 1\right)\right]}{\left[N\pi\left(\dfrac{\omega}{\omega_s} - 2m - 1\right)\right]^2} \equiv 4N\delta_N\left(\frac{\omega}{\omega_s} - 2m - 1\right),$$

$$(m = 0, 1, 2, \cdots), \tag{6.2.16}$$

where $\delta_N(x)$ is defined as

$$\delta_N(x) = N\left(\frac{\sin N\pi x}{N\pi x}\right)^2, \tag{6.2.17}$$

which reduces to the Dirac delta function for $N \to \infty$.

The final result for the frequency spectrum for synchrotron radiation from an electron passing through $2N$ pairs of permanent magnets per unit solid angle in the z direction takes the following form:

$$\frac{d^2F(\omega,\theta)}{d\Omega d\omega} = 4N\frac{3e^2}{4\pi^3}\sqrt{\frac{\mu_0}{\varepsilon_0}}\gamma^2\left(\frac{\omega}{\omega_{cr}}\right)^2 K_{2/3}^2\left(\frac{\omega}{\omega_{cr}}\right)\sum_m \delta_N\left(\frac{\omega}{\omega_s} - 2m - 1\right). \tag{6.2.18}$$

From (6.2.18), we find that the spectrum consists of fundamental and odd-order higher harmonic waves for large values of N. In addition, due to constructive interference, the power radiated from an electron passing through an array of $2N$ magnets per unit solid angle in the z direction is reinforced at the frequencies of the fundamental and higher harmonic waves by $4N^2$ times as compared with the power radiated from an electron bent by a single magnet. In Fig. 6.5 is illustrated the frequency spectrum corresponding to the pulse train in Fig. 6.4. Note that the envelope of spectral lines in Fig. 6.5 is proportional to the spectrum for the synchrotron radiation due to a single magnet given by (6.2.11). As seen from Fig. 6.5, the frequency spectrum for synchrotron radiation from an array of permanent magnets consists of fundamental and odd-order higher harmonic waves. For $\omega_{cr}/\omega_s \ll 1$ or $a_u \ll 1$, in particular, we can get something like a simple harmonic wave.

In the above discussion, we have treated an ideal case. However, there are actually some factors to blur the spectrum for synchrotron radiation. For example, in order to consider synchrotron radiation from a number of electrons, we must take into account the velocity distribution or kinetic energy distribution for electrons. If a group of electrons has a kinetic energy spread $m_0c^2\Delta\gamma$, we find from (6.2.10) that the corresponding observed frequency for the fundamental wave, for which $m = 0$ in (6.2.18), has a spread

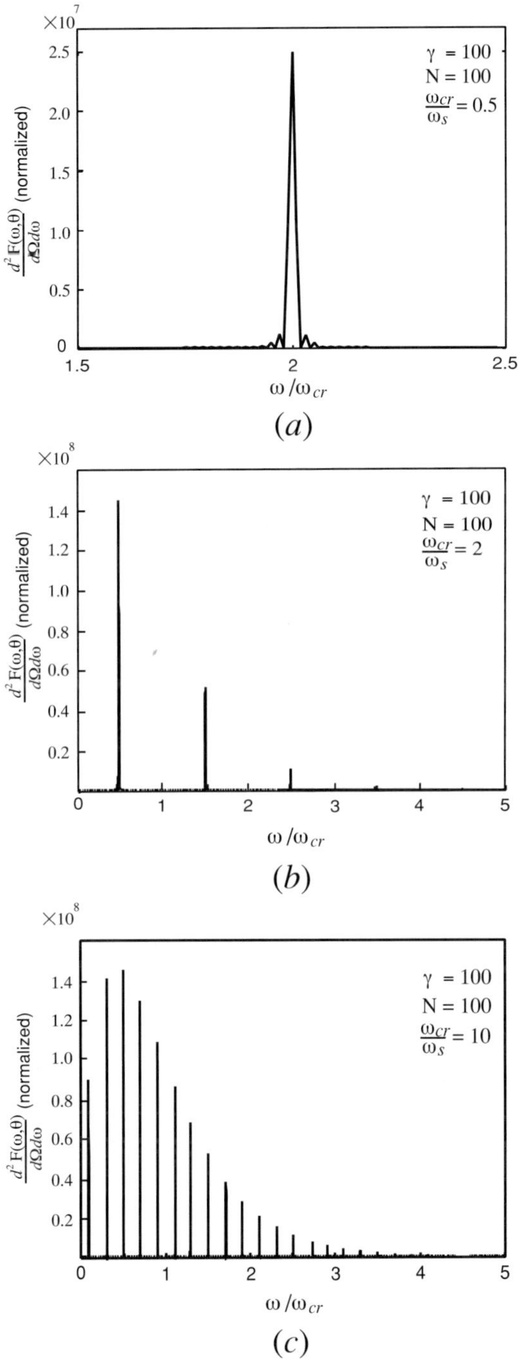

Fig. 6.5. Frequency spectrum for the pulse train in Fig. 6.4

$$\frac{\Delta\omega}{\omega_s} = 2\frac{\Delta\gamma}{\gamma}. \tag{6.2.19}$$

On the other hand, the fundamental frequency component in the spectrum (6.2.18) has a natural line width

$$\frac{\Delta\omega}{\omega_s} = \frac{1}{N}. \tag{6.2.20}$$

Hence, if synchrotron radiation from a group of electrons is to be as sharp as that from a single electron, we must have the following condition:

$$\frac{\Delta\gamma}{\gamma} << \frac{1}{N}. \tag{6.2.21}$$

Another factor to broaden the spectrum is a spread in the directions of electron velocities. If directions of electron velocities have a spread $\Delta\theta$, the corresponding observed frequency is found, from (6.2.10), to have a spread

$$\frac{\Delta\omega}{\omega_s} = \frac{\gamma^2}{1+(1/2)a_u^2}(\Delta\theta)^2, \tag{6.2.22}$$

where we have assumed $\Delta\theta << 1/\gamma$. For the frequency spread caused by the spread in the directions of electron velocities to be well within the natural line width, we need, from (6.2.20) and (6.2.22), the following condition:

$$(\Delta\theta)^2 << \frac{1+(1/2)a_u^2}{N\gamma^2}, \tag{6.2.23}$$

which is generally more difficult to satisfy than the condition (6.2.21). If the conditions (6.2.21) and (6.2.23) are not satisfied, we can not get the frequency spectrum as sharp as (6.2.18), but we only have a spectrum which is blurred, or in which sharp maxima are smoothed out.

6.3 Resonant Interaction of Electrons with Electromagnetic Wave

6.3.1 Condition for Resonant Interaction

Let us discuss quantitatively the interaction of an ensemble of electrons with a plane electromagnetic wave under the influence of a periodic transverse static magnetic field. The field components of a plane electromagnetic wave traveling in the z direction are expressed as

$$E_x = E_0 \sin(\omega t - kz + \phi), \tag{6.3.1}$$

$$B_y = \frac{E_0}{c} \sin(\omega t - kz + \phi), \tag{6.3.2}$$

where E_0 and ϕ denote a constant amplitude and phase of the electric field. From (6.2.5) and (6.3.2), we get a z component of the Lorentz force acting on the ensemble of electrons

$$\begin{aligned} -ev_x B_y &= -\frac{e^2 B_0 E_0}{\gamma m_0 k_0 c} \sin(\omega t - kz + \phi) \sin k_0 z \\ &= -\frac{e^2 B_0 E_0}{2\gamma m_0 k_0 c} \Big[\cos[\omega t - (k + k_0)z + \phi] - \cos[\omega t - (k - k_0)z + \phi]\Big]. \end{aligned} \tag{6.3.3}$$

The first term in the brackets on the right-hand side in (6.3.3) can be synchronized with the ensemble of electrons which is moving with a velocity less than that of light in vacuum. Specifically, if the condition

$$\frac{\omega}{k + k_0} = \bar{v}_z \tag{6.3.4}$$

is satisfied, the ensemble of electrons travels hand in hand with the beat wave produced by the coupling of a plane electromagnetic wave and periodic transverse static magnetic field. Note that the left-hand side of (6.3.4) represents the phase velocity of the beat wave. Then, the ensemble of electrons is subject to a consecutive influence of the longitudinal Lorentz force with the same sense. With the aid of (6.2.8), we can readily get the following resonance condition corresponding to (6.3.4):

$$\frac{\lambda_s}{\lambda_0} = \frac{1}{2\gamma^2}\left(1 + \frac{1}{2} a_u^2\right), \tag{6.3.5}$$

where λ_s represents the resonant wavelength of the electromagnetic wave. The resonant wavelength obtained above coincides with (6.2.9), with θ set equal to 0, which has apparently been found with the aid of a different kinematic method.

If the resonance or synchronism condition (6.3.4) is satisfied, the second term in the brackets on the right-hand side of (6.3.3) is smoothed out after the propagation of several beat wavelengths and only the first term, which is proportional to $\cos\phi$, survives. Then, in a similar fashion, we get from (2.10.5)

$$\frac{dK}{dt} = \frac{d}{dt}\left[m_0 c^2 (\gamma - 1)\right] = -ev_x E_x = -\frac{e^2 B_0 E_0}{2\gamma m_0 k_0} \cos\phi, \tag{6.3.6}$$

which describes the exchange of energy between the electron and the electromagnetic wave. In (6.3.6), K denotes the kinetic energy of the electron, and the phase ϕ can be considered to represent the position of the electron relative to the electric field. For $\cos\phi > 0$, we have $dK/dt < 0$, when the electromagnetic wave gets energy from the electron. On the other hand, for $\cos\phi < 0$, we have $dK/dt > 0$, when the electromagnetic wave loses energy to the electron. Furthermore, for $\cos\phi = 0$, we have $dK/dt = 0$, when there is no net energy transfer between the electron and the electromagnetic wave. Hence, if ϕ is uniformly distributed between 0 and 2π for the ensemble of electrons, we have the same amount of accelerated and decelerated electrons, and then there is no net energy transfer between the ensemble of electrons and the electromagnetic wave.

6.3.2 Small Signal Gain

We now investigate the energy exchange between the electron and the electromagnetic wave around the resonant frequency. Starting from (2.10.5), we calculate the overall energy transfer from the ensemble of electrons to the electromagnetic wave around the resonant frequency. Specifically, we apply the energy exchange relation (6.3.6) to a more general case which describes the electron-wave interaction in the vicinity of resonance. First, we rewrite (6.3.6) as

$$\frac{dK}{dt} = -ev_x[z(t)]E_x[z(t),t], \tag{6.3.7}$$

where we have made it explicit that the values of fields at the position of the electron at the time t should be substituted on the right-hand side. Substituting (6.2.5) and (6.3.1), and extracting only the term relevant to the resonant interaction, we get

$$\frac{dK}{dt} = -\frac{e^2 B_0 E_0}{2\gamma\, m_0 k_0} \cos\left[\omega t - (k + k_0) z(t) + \phi\right]. \tag{6.3.8}$$

In order to express the coordinate of the electron z explicitly in terms of the time t, we solve the following equation of motion for the electron, which can be obtained from (2.10.6):

$$\frac{dv_z}{dt} = \frac{e}{\gamma\, m_0 c^2}\left(v_x E_x\right) v_z - \frac{e}{\gamma\, m_0} v_x B_y. \tag{6.3.9}$$

Inserting (6.2.5), (6.3.1), and (6.3.2), and using the relation $z = \bar{v}_z t$ as the zeroth-order approximation, we find

$$\frac{dv_z}{dt} = -\frac{e^2 B_0 E_0}{2\gamma^2 m_0^2 k_0 c}\left(1 - \frac{\bar{v}_z}{c}\right) \cos\left[\omega t - (k + k_0)\bar{v}_z t + \phi\right]. \tag{6.3.10}$$

Rewriting (6.3.10) with the aid of (6.2.8) yields

$$\frac{dv_z}{dt} = -\frac{e^2 B_0 E_0}{2\gamma^2 m_0^2 k_0 c} \Gamma \cos(\Omega t + \phi), \tag{6.3.11}$$

with

$$\Omega = \Gamma\omega - \omega_0, \quad \Gamma = \frac{1}{2\gamma^2}\left(1 + \frac{1}{2}a_u^2\right), \tag{6.3.12}$$

where $\Omega = 0$ defines the resonant frequency ω_s.

Integrating (6.3.11) from 0 to t with respect to time, we get

$$v_z(t) = \frac{dz}{dt} = \bar{v}_z + \Delta v_z(t), \tag{6.3.13}$$

where

$$\Delta v_z(t) = -\int_0^t \frac{e^2 B_0 E_0}{2\gamma^2 m_0^2 k_0 c} \Gamma \cos(\Omega t' + \phi) dt'$$

$$= -\frac{e^2 B_0 E_0}{2\gamma^2 m_0^2 k_0 c\Omega} \Gamma\left[\sin(\Omega t + \phi) - \sin\phi\right]. \tag{6.3.14}$$

Then, integration of (6.3.13) with respect to time gives

$$z(t) = \bar{v}_z t + \Delta z(t), \tag{6.3.15}$$

where

$$\Delta z(t) = \frac{e^2 B_0 E_0}{2\gamma^2 m_0^2 k_0 c\Omega^2} \Gamma\left[\cos(\Omega t + \phi) - \cos\phi + \Omega t \sin\phi\right]. \tag{6.3.16}$$

We find that the position or the phase of the electron slips relative to the electric field as a result of the electron-wave interaction. Inserting (6.3.15), together with (6.3.16), in (6.3.8), we get the following result:

$$\frac{dK}{dt} = -\frac{e^2 B_0 E_0}{2\gamma m_0 k_0} \cos\left[\Omega t + \phi + \Delta\phi(t)\right], \tag{6.3.17}$$

where

$$\Delta\phi(t) = -\frac{e^2 B_0 E_0}{2\gamma^2 m_0^2 c\Omega^2}\left[\cos(\Omega t + \phi) - \cos\phi + \Omega t \sin\phi\right]. \tag{6.3.18}$$

For $\Delta\phi << \pi$, which is valid for small values of E_0, Eq. (6.3.17) reduces to

$$\frac{dK}{dt} = -\frac{e^2 B_0 E_0}{2\gamma m_0 k_0}\cos(\Omega t + \phi)$$

$$-\frac{\left(e^2 B_0 E_0\right)^2}{4\gamma^3 m_0^3 k_0 c \Omega^2}\sin(\Omega t + \phi)\left[\cos(\Omega t + \phi) - \cos\phi + \Omega t \sin\phi\right]. \qquad (6.3.19)$$

Assume that the ensemble of electrons is distributed uniformly from 0 to 2π in the phase of the beat wave. Then, averaging (6.3.19) with respect to ϕ over the ensemble of uniformly distributed electrons, we get

$$\left\langle \frac{dK}{dt} \right\rangle_\phi = \frac{1}{2\pi}\int_0^{2\pi} \frac{dK}{dt}\, d\phi = \frac{\left(e^2 B_0 E_0\right)^2}{8\gamma^3 m_0^3 k_0 c \Omega^2}\left(\sin \Omega t - \Omega t \cos \Omega t\right), \qquad (6.3.20)$$

which represents the average energy lost or gained by a single electron per unit time. Integrating (6.3.20) over the interaction time τ defined by $\tau = L / \bar{v}_z$, where L denotes the interaction distance, we obtain for the average energy lost or gained by a single electron during the time of electron-wave interaction,

$$\langle K \rangle_\phi = \int_0^\tau \left\langle \frac{dK}{dt'} \right\rangle_\phi dt' = \frac{\left(e^2 B_0 E_0\right)^2}{8\gamma^3 m_0^3 k_0 c}\tau^3 g(\Omega\tau), \qquad (6.3.21)$$

where $g(\Omega\tau)$ is a gain function defined in (5.6.12) as

$$g(\Omega\tau) = \frac{2(1 - \cos \Omega\tau) - \Omega\tau \sin \Omega\tau}{(\Omega\tau)^3}, \qquad (6.3.22)$$

with Ω given by (6.3.12) for the present case, which is defined differently from the previous case.

The power gain for the electromagnetic wave per one transit of the interaction distance L can be calculated with the aid of (6.3.21). The average power increment $<\Delta P>$ for the electromagnetic wave corresponding to the average power loss for the ensemble of electrons per one transit can be expressed as

$$\langle \Delta P \rangle = -\frac{I}{e}\langle \Delta K \rangle_\phi$$

$$= -I \frac{e^3 (E_0 B_0)^2}{8\gamma^3 m_0^3 k_0 c}\tau^3 g(\Omega\tau), \qquad (6.3.23)$$

where I denotes the average beam current, and I/e corresponds to the number of electrons which traverse a plane perpendicular to the electron beam per unit time. On the other hand, the time-average effective power P for the electromagnetic wave passing through an area A perpendicular to the electron beam is given by

$$P = \left(\frac{1}{2} \sqrt{\frac{\varepsilon_0}{\mu_0}} E_0^2 \right) cA = \frac{1}{2\mu_0} E_0^2 A, \tag{6.3.24}$$

where we assume that A is greater than the cross-sectional area of the electron beam. From (6.3.23) and (6.3.24), the small-signal gain G for the electromagnetic wave in one transit of the interaction distance can be represented as

$$G = \frac{\langle \Delta P \rangle}{P} = -I \frac{\mu_0 e^3 B_0^2}{4\gamma^3 m_0^3 k_0 cA} \tau^3 g(\Omega\tau). \tag{6.3.25}$$

The main gain is obtained in the region $\Omega\tau < 0$ where $g(\Omega\tau) < 0$, and the power gain G becomes maximum at $\Omega\tau = -2.6$. Referring to (6.3.12) and (6.3.5), we see that we get the maximum gain for the frequency slightly less than the resonant frequency ω_s.

Finally, we discuss the gain bandwidth within which we can have values of the power gain greater than half the maximum gain. Referring to Fig. 5.13, we see that the full width at half maximum for the gain function $g(\Omega\tau)$ is nearly equal to π, namely, $\Delta(\Omega\tau) = \pi$, which is written with the aid of (6.3.12) as

$$(\Delta\omega)\Gamma\tau = \pi. \tag{6.3.26}$$

This relation can be further rewritten in the form

$$\frac{\Delta\omega}{\omega} = \frac{\bar{v}_z}{L} \frac{1}{\omega\Gamma} \pi = \frac{1}{L} \frac{\bar{v}_z}{\omega_0} \pi = \frac{1}{L} \frac{\pi}{k_0} = \frac{\lambda_0}{2L}. \tag{6.3.27}$$

If we assume that the interaction distance L is equal to N times the undulator period, namely, $L = N\lambda_0$, we finally get for the gain bandwidth,

$$\frac{\Delta\omega}{\omega} = \frac{1}{2N}. \tag{6.3.28}$$

In Fig. 6.6 (a), we show the laser spectrum obtained in the first operation of a free-electron laser oscillator at Stanford [6.8], which used a superconducting double helix to generate a transverse periodic static magnetic field. The corresponding synchrotron radiation spectrum is also shown in Fig. 6.6 (b). In Fig. 6.6 (b), we do not see sharp line spectra as predicted in the foregoing discussion for the synchrotron radiation, but only a smoothed spectrum with four clear peaks. This is because the spectrum in Fig. 6.6 (b) is not due to the synchrotron radiation by a single electron but due to that emitted by large numbers of electrons with a velocity or energy spread. From these figures, we also find that the peak of the spectrum for the laser oscillation appears at a wavelength slightly longer than that for the corresponding synchrotron radiation, as clarified in the foregoing discussion. Further-

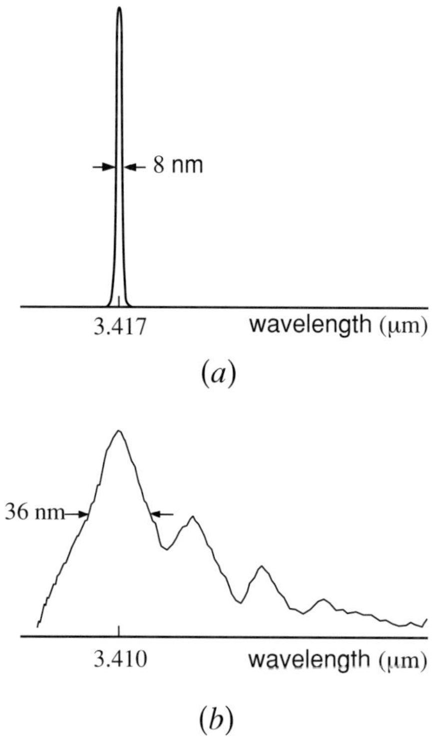

Fig. 6.6. Laser spectrum obtained in the first operation of a free-electron laser oscillator at Stanford (*a*), and the corresponding spectrum for synchrotron radiation (*b*)[6.8]

more, it should be noted that the spectra in Figs. 6.6 (*a*) and (*b*) are expressed on different scales and the peak value of the spectrum for the laser oscillation is actually 10^8 times as large as that for the corresponding synchrotron radiation. The reason why such a sharp spectrum with an enormous power is obtained in the laser oscillation is that it has been produced by the multiple interaction of bunched electrons and optical pulses multiply reflected back and forth between two reflecting mirrors composing a Fabry-Perot resonator (see Fig. 6.1).

6.4 Trapping of Electrons in the Beat Wave

As the electron-wave interaction progresses and the electromagnetic wave grows substantially, electrons are trapped in the beat wave produced by the coupling of the electromagnetic wave and the transverse periodic static magnetic field. The basic equation for discussing electron trapping in the beat wave is obtained from (6.3.8), which is reproduced here,

$$\frac{dK}{dt} = -\frac{e^2 B_0 E_0}{2\gamma m_0 k_0} \cos\left[\omega t - (k + k_0)z + \phi\right]. \tag{6.4.1}$$

Introducing a new phase variable defined by

$$\psi = (k + k_0)z - \omega t - \phi + \frac{\pi}{2}, \tag{6.4.2}$$

we get

$$\frac{dK}{dt} = -\frac{e^2 B_0 E_0}{2\gamma m_0 k_0} \sin\psi . \tag{6.4.3}$$

Differentiating (6.4.2) with respect to z leads to

$$\frac{d\psi}{dz} = k + k_0 - \frac{\omega}{\bar{v}_z} = k_0 + \left(1 - \frac{c}{\bar{v}_z}\right)k , \tag{6.4.4}$$

where we have used $dz / dt \approx \bar{v}_z$, which can be justified by (6.3.13) and (6.3.14). With the aid of (6.2.8), we can rewrite (6.4.4) as

$$\frac{d\psi}{dz} = k_0 - \frac{k}{2\gamma^2}\left(1 + \frac{1}{2}a_u^2\right). \tag{6.4.5}$$

At resonance, $d\psi/dz$ vanishes. Then, representing the value of γ at resonance as γ_r, which satisfies

$$k_0 = \frac{k}{2\gamma_r^2}\left(1 + \frac{1}{2}a_u^2\right), \tag{6.4.6}$$

we have

$$\frac{d\psi}{dz} = k_0\left[1 - \left(\frac{\gamma_r}{\gamma}\right)^2\right]. \tag{6.4.7}$$

If we put

$$\gamma = \gamma_r + \Delta\gamma , \quad \Delta\gamma << \gamma_r , \tag{6.4.8}$$

Equation (6.4.7) reduces to

$$\frac{d\psi}{dz} = 2k_0 \frac{\Delta\gamma}{\gamma_r}, \tag{6.4.9}$$

where, for simplicity, γ_r is assumed to be constant.

Next, in order to get the second derivative of ψ, we rewrite (6.4.3) as

$$\frac{dK}{dz} = \frac{dK}{dt}\frac{1}{(dz/dt)} = -\frac{e^2 B_0 E_0}{2\gamma\, m_0 k_0 c}\sin\psi\,, \tag{6.4.10}$$

where we have assumed $v_z \approx c$. Here, noting the relation

$$K = m_0 c^2(\gamma - 1), \tag{6.4.11}$$

together with (6.4.8), we find from (6.4.10),

$$\frac{d}{dz}(\Delta\gamma) = -\frac{e^2 B_0 E_0}{2\gamma_r m_0^2 k_0 c^3}\sin\psi\,. \tag{6.4.12}$$

Then, from (6.4.9), we get

$$\frac{d^2\psi}{dz^2} = \frac{2k_0}{\gamma_r}\frac{d}{dz}(\Delta\gamma) = -k_\psi^2 \sin\psi\,, \tag{6.4.13}$$

where

$$k_\psi^2 = \frac{e^2 B_0 E_0}{\gamma_r^2 m_0^2 c^3}\,. \tag{6.4.14}$$

On the basis of (6.4.13), together with the relation (6.4.9), we can discuss the nonlinear behavior of electrons in the phase of the beat wave. For a detailed analytical treatment of (6.4.13), refer to the discussion in Sect. 5.7.

Let us consider a typical case where the parameters of the undulator (λ_0 and B_0) are constant and the wave field has a sufficiently large amplitude, varying only slightly with z. Then, we examine how much energy we can extract from the ensemble of electrons with initial kinetic energy $m_0c^2(\gamma_i - 1)$, which are uniformly distributed in the phase of the beat wave at the initial state. In order to extract energy efficiently from the ensemble of electrons, we must allow it to be captured in the beat wave. Then, from (6.4.9) and (5.7.13), γ_i must satisfy the following condition:

$$0 < \gamma_i - \gamma_r \le (\Delta\gamma)_m, \tag{6.4.15}$$

where

$$(\Delta\gamma)_m = \gamma_r \frac{k_\psi}{k_0}, \tag{6.4.16}$$

which is the value of $\Delta\gamma$ corresponding to the height or maximum value of the separatrix. In view of (6.3.12), the gain function $g(\Omega\tau)$ may be regarded as a function of γ_i when the frequency of electromagnetic wave is kept constant (see

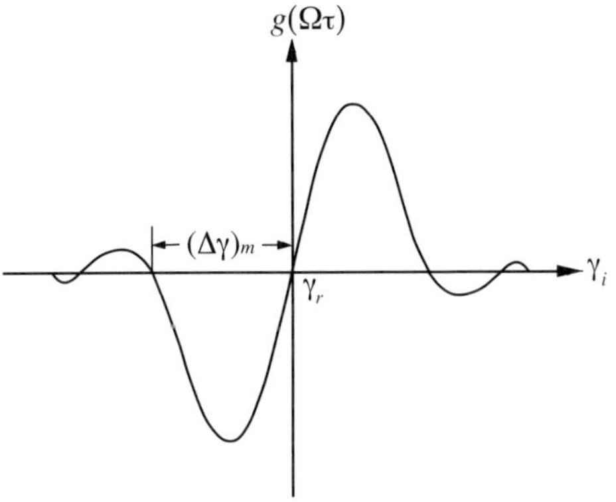

Fig. 6.7. Gain function regarded as a function of γ_i for the case where the frequency of electromagnetic wave is kept constant

Fig. 6.7). Then, for the value of γ_i in (6.4.15), we get a negative value for gain function $g(\Omega\tau)$ or a positive value for small-signal gain G. When γ_i satisfies (6.4.15) and the ensemble of electrons undergoes about a half cycle of orbital motion in the phase space of the beat wave, the growth of the electromagnetic wave gets to saturation with maximum energy extracted from it (see Fig. 5.17). Then, we have the length of the undulator equal to

$$L = \frac{\pi}{k_\psi} = N\lambda_0. \tag{6.4.17}$$

where N denotes the number of undulator periods.

With the aid of (6.4.16) and (6.4.17), we can get the efficiency of energy transfer from the ensemble of electrons to the electromagnetic wave. The efficiency of energy transfer η is defined as the ratio of the maximum energy lost by the ensemble of electrons to its initial kinetic energy,

$$\eta = \frac{m_0 c^2 (\Delta\gamma)_m}{m_0 c^2 (\gamma_i - 1)} \approx \frac{(\Delta\gamma)_m}{\gamma_r}. \tag{6.4.18}$$

where we have assumed $\gamma_i >> 1$ and $\gamma_i \approx \gamma_r$. Using the relations (6.4.16) and (6.4.17), we get

$$\eta \approx \frac{1}{2N}. \tag{6.4.19}$$

Since N is generally a large number in the range 50~100, the efficiency is rather low for the mode of laser operation considered in the above example. In addition, after the efficiency of energy transfer (6.4.19) has been attained, initially mono-energetic electrons have got energy spread corresponding to (6.4.18). It should also be noted that we cannot expect low efficiency to be improved by decreasing the number of undulator periods N. As is evident from (6.2.20), the frequency spectrum of the electromagnetic wave radiated is broadened in proportion to $1/N$ as N decreases. Hence the decrease in N causes, at the same time, the coherence of the radiation to be degraded.

In the above example, we have assumed that γ_r is constant. For this case, we cannot expect high efficiency of energy transfer from the ensemble of electrons to the electromagnetic wave, as seen from (6.4.19). In order to get over this difficulty of low efficiency, we must somehow maintain the resonant electron-wave interaction after it has reached saturation with the energy transfer efficiency (6.4.19). A typical remedy for it is to reduce the value of γ_r adiabatically in accordance with the decrease in the mean value of γ, which is caused by energy transfer from the ensemble of electrons to the electromagnetic wave. From (6.4.6) and (6.2.7), we have

$$\gamma_r(z) = \left[\frac{k}{2k_0(z)}\left[1 + \frac{1}{2}a_u^2(z)\right]\right]^{1/2}$$

$$= \left[\frac{\lambda_0(z)}{2\lambda}\left[1 + \frac{1}{2}\left[\frac{e\lambda_0(z)B_0(z)}{2\pi m_0 c}\right]^2\right]\right]^{1/2}, \tag{6.4.20}$$

where the value of λ is kept constant for the present mode of laser operation. Then, if we change one or both of the values of undulator parameters λ_0 and B_0 in the longitudinal direction adiabatically in accordance with the decrease in the mean value of γ, the value of γ_r can be changed accordingly. Thus we can maintain the resonant electron-wave interaction, and substantially improve the efficiency of energy transfer between them. Although efficiency enhancement in the free-electron laser is an important subject, a more detailed discussion of this subject is beyond the scope of this book, which is devoted only to the basics of electron-wave interaction in relativistic electron beams. For a more elaborate single-particle treatment of the free-electron laser, the reader will be recommended to refer to [6.12].

7. Collective Theory of the Free-Electron Laser

7.1 Introduction

The free-electron laser has the great advantages of high output power, continuous tunability over wide frequency range, and possible high efficiency of energy transfer from the kinetic energy of a relativistic electron beam to the electromagnetic wave energy. The amplification mechanism for the free-electron laser can be explained on the basis of the models of stimulated scattering of electromagnetic waves by a relativistic electron beam, namely, the stimulated Compton scattering in the shorter wavelength region and the stimulated Raman scattering in the longer wavelength region. The former scattering process treats the scattering by individual electrons composing a relativistic electron beam, while the latter scattering process considers the scattering by collective oscillation of electrons excited in the electron beam. In this chapter, we discuss the amplification mechanism for the free-electron laser in the longer wavelength region of millimeter to submillimeter waves, with the aid of the stimulated Raman scattering model. From a general point of view, the process of the stimulated Raman scattering in a relativistic electron beam can be regarded as a parametric interaction of the three waves, i.e., the pump wave, the scattered wave (positive-energy wave), and the electron plasma wave (negative-energy wave). In other words, under the influence of the pump wave, energy is exchanged between the scattered wave and the electron plasma wave, of which the former is extracted as the laser output. For the pump wave, an intense electromagnetic wave is used. However, in place of an electromagnetic wave pump, we can also use a transverse static magnetic field varied periodically in the direction of beam flow to get a laser action. This is because a transverse periodic static magnetic field behaves in the rest frame of the electron beam as an electromagnetic wave varying temporally with constant frequency, which plays the same role as the pump wave.

This chapter clarifies the amplification mechanism for a long-wavelength free-electron laser using an intense electromagnetic wave as the pump wave, on the basis of coupled-mode theory. In the first part of the following discussion, we derive the coupled-mode equations relating the pump wave, the scattered wave, and the electron plasma wave propagated in a relativistic electron beam, with the aid of the Maxwell equations, the relativistic equation of motion for the electron, and the equation of continuity for the flow of electrons. For the special case of linear regime, we can get straightforward the spatial growth rate for the scattered wave and the electron plasma wave from these coupled-mode equations. For the case where the linear approximation does not hold, we solve the coupled-mode equations to get analytical solutions for three waves. Then, we formulate the energy relations for three waves, discussing energy exchange between the scattered wave and the electron plasma wave under the influence of the pump wave. Finally, we investigate the saturation mechanism for the laser output due to the trapping of electrons in the electron plasma wave, and the efficiency of energy transfer from the kinetic energy of a relativistic electron beam to the laser output.

7.2 Stimulated Raman Scattering in a Relativistic Electron Beam

As mentioned earlier, the long-wavelength free-electron laser extracts its output energy from the kinetic energy of a relativistic electron beam, through the process of the stimulated Raman scattering in it, or through the parametric interaction of three waves, i.e., the pump wave, the scattered wave (positive-energy wave), and the electron plasma wave (negative-energy wave). Figure 7.1 shows two typical models of the Raman-regime free-electron laser. In the model shown in Fig. 7.1 (a), we use, as the pump wave, an intense electromagnetic wave propagated against the flow of the electron beam. In the model shown in Fig. 7.1 (b), on the other hand, a transverse static magnetic field varied periodically along the electron beam acts as the pump wave. For the case of Fig. 7.1 (b), a periodic static magnetic field in the laboratory frame behaves as an electromagnetic wave varying temporally with constant frequency in the rest frame of the electron beam, which plays the same role as the pump wave in Fig. 7.1 (a). Hence, in the laboratory frame, we may regard a transverse static magnetic field with periodic variation in the longitudinal direction as a pump wave with zero frequency. Thus, viewed in the rest frame of the electron beam, the pump wave is scattered by the electron plasma oscillation, for either case of Fig. 7.1 (a) or Fig. 7.1 (b). Then, if we observe the backscattered wave propagated parallel to the flow of the electron beam in the laboratory frame, its frequency becomes higher than that of the pump wave due to the Doppler shift.

For an electromagnetic wave pump with frequency ω_i and wave number k_i (< 0) in the laboratory frame, let us find the frequency of the backscattered wave, with the aid of the Lorentz transformation for frequency and wave numbers given by (2.5.5). First, the frequency of the pump wave in the rest frame of the electron beam, ω_i' becomes

$$\omega_i' = \gamma_0(\omega_i - v_0 k_i), \tag{7.2.1}$$

with

$$\gamma_0 = \frac{1}{\sqrt{1-\beta_0^2}}, \quad \beta_0 = \frac{v_0}{c}, \tag{7.2.2}$$

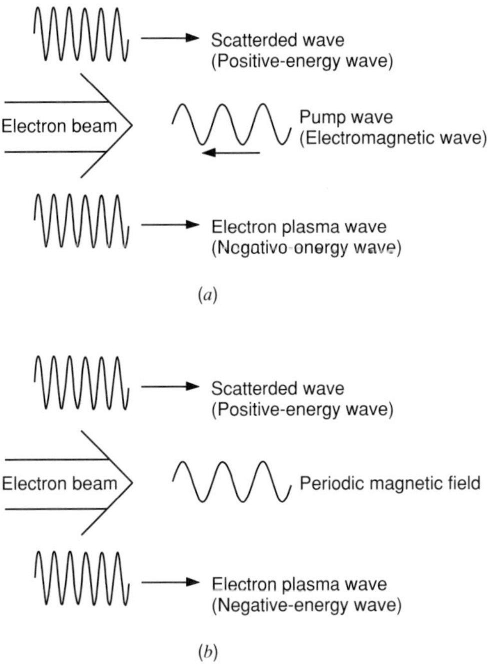

Fig. 7.1. Models of Raman-regime free-electron lasers

where v_0 denotes the drift velocity of the electron beam. Inserting in (7.2.1) the relation

$$k_i = -\frac{\omega_i}{c}\sqrt{\varepsilon_i}\,, \quad \varepsilon_i = 1 - \left(\frac{\omega_p}{\omega_i}\right)^2, \tag{7.2.3}$$

ω_p being the electron plasma frequency for the electron beam, we get for the frequency of the pump wave in the rest frame of the electron beam,

$$\omega_i' = \gamma_0\left(1 + \beta_0\sqrt{\varepsilon_i}\right)\omega_i. \tag{7.2.4}$$

In the rest frame of the electron beam, the electron plasma oscillation at the frequency $\omega_l' = \omega_p$ will be excited by the pump wave of frequency ω_i', and it will scatter the pump wave. This process corresponds to Raman scattering in the electron plasma at rest. As a result, the frequency of the backscattered wave, ω_s' becomes smaller than ω_i',

$$\omega_s' = \omega_i' - \omega_p. \tag{7.2.5}$$

The frequency of the backscattered wave in the laboratory frame, ω_s can be obtained by transforming ω_s' back to the laboratory frame as

$$\omega_s = \gamma\left(\omega_s' + v_0 k_s'\right), \tag{7.2.6}$$

where k_s' is the wave number corresponding to ω_s',

$$k_s' = \frac{\omega_s'}{c}\sqrt{\varepsilon_s'}, \quad \varepsilon_s' = 1 - \left(\frac{\omega_p}{\omega_s'}\right)^2. \tag{7.2.7}$$

Substituting (7.2.7), together with (7.2.5) and (7.2.4), in (7.2.6), we can get for ω_s,

$$\frac{\omega_s}{\omega_i} = \gamma_0^2\left[1 + \beta_0\sqrt{\varepsilon_i} - \frac{\omega_p}{\gamma_0\omega_i} + \beta_0\sqrt{\left(1 + \beta_0\sqrt{\varepsilon_i}\right)\left(1 + \beta_0\sqrt{\varepsilon_i} - \frac{2\omega_p}{\gamma_0\omega_i}\right)}\right]. \tag{7.2.8}$$

For the special case of $\beta_0 \approx 1$ and $\omega_p << \omega_i$, Eq. (7.2.8) reduces to

$$\omega_s = 4\gamma_0^2\omega_i. \tag{7.2.9}$$

Similarly, for a static magnetic field pump with the spatial period λ_i, we can obtain for the frequency of the backscattered wave in the laboratory frame,

$$\frac{\omega_s}{c|k_i|} = \gamma_0^2\left[\beta_0 - \frac{\omega_p}{\gamma_0 c|k_i|} + \beta_0\sqrt{\beta_0\left[\beta_0 - \frac{2\omega_p}{\gamma_0 c|k_i|}\right]}\right], \tag{7.2.10}$$

which reduces, for an extremely relativistic case, to

$$\omega_s = 2\gamma_0^2 c|k_i|. \tag{7.2.11}$$

As seen from the above expression for ω_s, the frequency of the backscattered wave in the laboratory frame becomes extremely high as the drift velocity of the electron beam approaches the velocity of light in vacuum. In this regard, refer to

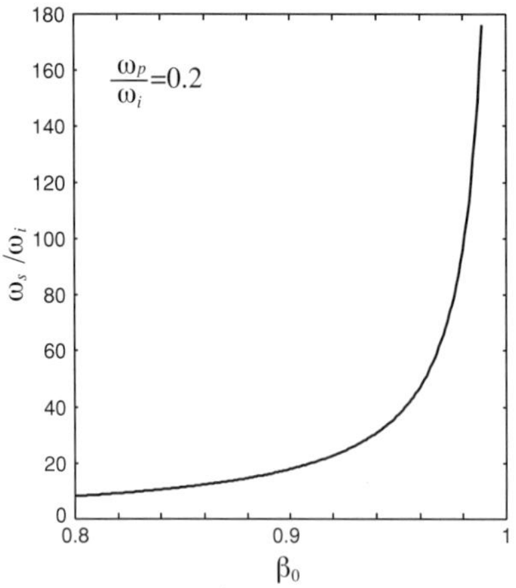

Fig. 7.2. Frequency of the back-scattered wave versus the drift velocity of the electron beam

Fig. 7.2 in which the frequency of the backscattered wave versus the drift velocity of the electron beam is shown according to (7.2.8). Thus, if a microwave is used as the pump wave, we can easily get a scattered wave in the millimeter or submillimeter wave region.

In the models shown in Fig. 7.1, the transverse scattered wave, which is a positive-energy wave, is coupled to the longitudinal electron plasma wave in a relativistic electron beam, which is a negative-energy wave, under the influence of the pump wave. And then, energy is exchanged between the scattered wave and the electron plasma wave. In this way, the free-electron laser extracts its output energy not from the pump wave but from the kinetic energy of the electron beam. Thus, from the viewpoint of the mechanism for energy transfer, the free-electron laser is found to be analogous to a microwave tube such as the traveling wave tube. However, it should be noted that the traveling wave tube extracts energy through the direct coupling of electron beam and electromagnetic waves. In the free-electron laser, on the other hand, energy is transferred from the electron beam to the electromagnetic wave with the help of the pump wave.

7.3 Basic Equations

The basic equations for the subsequent analysis are the Maxwell equations, the relativistic equation of motion for the electron, and the equation of continuity for the flow of electrons. In these basis equations, we must take into account physical quantities to the second order with respect to oscillating field components. For simplicity, let us assume that physical quantities treated in the following discussion are varied only in the direction of beam flow, i.e., in the z direction but uniform in the transverse direction. Then, the Maxwell equations can be separated into the equations for the electromagnetic wave mode and those for the electron plasma wave mode. First, in the Cartesian coordinate system, the electromagnetic wave mode can be represented in terms of fields E_x, B_y, and current density J_x, which satisfy the following equations:

$$\frac{\partial E_x}{\partial z} = -\frac{\partial B_y}{\partial t}, \quad -\frac{1}{\mu_0}\frac{\partial B_y}{\partial z} = \varepsilon_0 \frac{\partial E_x}{\partial t} + J_x, \tag{7.3.1}$$

$$J_x = -(n_0 + n)ev_x, \tag{7.3.2}$$

where n_0 denotes the average number density of electrons, n and v_x being the oscillating parts in the number density of electrons and the x component of the electron velocity.

On the other hand, the electron plasma wave mode can be expressed in terms of field E_z, and current and charge densities J_z, ρ, which are governed by the following equations:

$$\varepsilon_0 \frac{\partial E_z}{\partial t} + J_z = 0, \qquad \frac{\partial E_z}{\partial z} = \frac{\rho}{\varepsilon_0}, \tag{7.3.3}$$

$$J_z = -(n_0 v_z + n v_0 + n v_z)e, \qquad \rho = -ne, \tag{7.3.4}$$

where v_z is the oscillating part in the z component of the electron velocity.

In addition, if we assume that the variation in the electron velocity is much less than its drift velocity, we obtain, from (2.10.6), the following expressions for the relativistic equation of motion for the electron, which are correct to the second order in terms of oscillating field components:

$$\frac{\partial v_x}{\partial t} + v_0 \frac{\partial v_x}{\partial z} = -\frac{e}{\gamma_0 m_0}\left(E_x - v_0 B_y\right) + \frac{e}{\gamma_0 m_0}\left(v_z B_y + \frac{\beta_0}{c} v_x E_z\right)$$
$$-v_z \frac{\partial v_x}{\partial z} - \gamma_0^2 \frac{\beta_0}{c} v_z \left(\frac{\partial v_x}{\partial t} + v_0 \frac{\partial v_x}{\partial z}\right), \tag{7.3.5}$$

$$\frac{\partial v_z}{\partial t} + v_0 \frac{\partial v_z}{\partial z} = -\frac{e}{\gamma_0^3 m_0} E_z - \frac{e}{\gamma m_0} v_x B_y + \frac{e}{\gamma m_0} \frac{\beta_0}{c}\left(v_x E_x + 2 v_z E_z\right)$$
$$-v_z \frac{\partial v_z}{\partial z} - \gamma_0^2 \frac{\beta_0}{c} v_z \left(\frac{\partial v_z}{\partial t} + v_0 \frac{\partial v_z}{\partial z}\right), \tag{7.3.6}$$

In deriving (7.3.5) and (7.3.6), we have assumed $\gamma_0^2 \beta_0 (v_z / c) << 1$, which is a valid approximation for the Raman-regime free-electron laser.

Finally, the equation of continuity for the flow of electrons takes the form

$$\frac{\partial n}{\partial t} + v_0 \frac{\partial n}{\partial z} + n_0 \frac{\partial v_z}{\partial z} = -\frac{\partial}{\partial z}(n v_z). \tag{7.3.7}$$

Equations (7.3.1) to (7.3.7) are the basic equations for the subsequent analysis.

7.4 Coupled-Mode Equations

In the absence of the pump wave, the electromagnetic and electron plasma wave modes in the relativistic electron beam can be propagated independently of each other. In the presence of the pump wave, however, these wave modes are coupled to each other, through terms of the second order with respect to oscillating field components in the basic equations, as a result of which energy is exchanged between them. In order to help the audience to comprehend the concept of the parametric interaction among the three waves, we show in Fig. 7.3 how the pump wave, the scattered wave (positive-energy wave), and the electron plasma wave (negative-energy wave) are coupled with each other on the dispersion diagram. In Fig. 7.3, the curve 1 expresses the dispersion relation for the pump wave and the scattered wave, and the straight lines 2 and 3 those for the electron plasma wave. The posi-

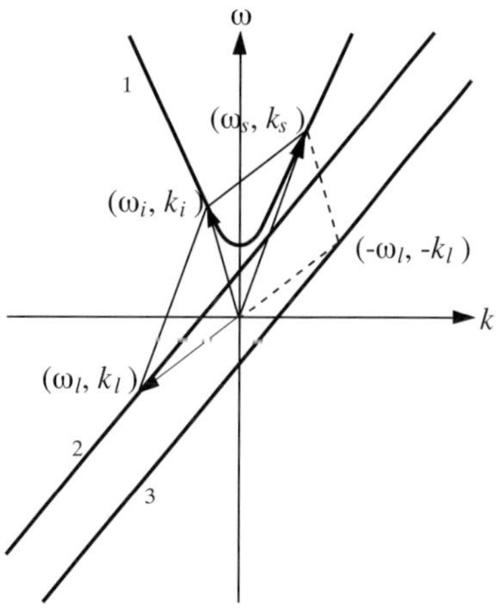

Fig. 7.3. Parametric interaction of pump, scattered and electron plasma waves

tive ω part of the line 2 and the negative ω part of the line 3 correspond to the positive-energy wave, while the negative ω part of the line 2 and the positive ω part of the line 3 represent the negative-energy wave.

Now, let the frequencies of the pump, scattered and electron plasma waves be denoted by ω_i, ω_s, and ω_l, and the corresponding wave numbers by k_i, k_s, and k_l. Then, we have the following phase-matching conditions:

$$\omega_s + \omega_l = \omega_i, \quad k_s + k_l = k_i, \tag{7.4.1}$$

with

$$k_s = \frac{\omega_s}{c}\sqrt{\varepsilon_s}, \quad \varepsilon_s = 1 - \left(\frac{\omega_p}{\omega_s}\right)^2, \quad \omega_l - v_0 k_l = \frac{\omega_p}{\gamma_0}, \tag{7.4.2}$$

where ω_s is given by (7.2.8). In (7.4.1), we assume $\omega_l < 0$. This assumption is mathematically acceptable since our analysis is carried out in the complex domain. However, it should be noted that the actual frequency of the electron plasma wave to be observed is given by $|\omega_l|$.

In the following discussion, we will try to find the coupled-mode equations for relating the pump, scattered and electron plasma waves on the assumption of weak coupling among the three waves. For this purpose, we express the field components for the electromagnetic waves as the sums of those for the pump and scattered waves,

$$E_x = \frac{1}{2}\left[\tilde{E}_i e^{j(\omega_i t - k_i z)} + \tilde{E}_s e^{j(\omega_s t - k_s z)} + c.c.\right], \tag{7.4.3}$$

$$B_y = \frac{1}{2}\left[\tilde{B}_i e^{j(\omega_i t - k_i z)} + \tilde{B}_s e^{j(\omega_s t - k_s z)} + c.c.\right], \tag{7.4.4}$$

$$v_x = \frac{1}{2}\left[\tilde{v}_i e^{j(\omega_i t - k_i z)} + \tilde{v}_s e^{j(\omega_s t - k_s z)} + c.c.\right], \tag{7.4.5}$$

where the subscripts i and s refer to the pump and scattered waves, respectively, and c.c. denotes the complex conjugates of the preceding functions. In (7.4.3) ~ (7.4.5), we assume that the amplitudes of the pump and scattered waves vary slowly in terms of z and t with the condition

$$\left|\frac{\partial \tilde{A}_m}{\partial z}\right| << \left|k_m \tilde{A}_m\right|, \quad \left|\frac{\partial \tilde{A}_m}{\partial t}\right| << \left|\omega_m \tilde{A}_m\right|, \quad (m = i, s), \tag{7.4.6}$$

where $\tilde{A}_m$ denotes the amplitude of any field component for the pump and scattered waves.

For the first-order quantities with respect to the pump and scattered waves, we have the following amplitude relations:

$$\tilde{B}_s = \frac{k_s}{\omega_s}\tilde{E}_s, \quad \tilde{v}_s = j\frac{e}{\gamma_0 m_0 \omega_s}\tilde{E}_s, \quad \tilde{v}_i = j\frac{e}{\gamma_0 m_0 \omega_i}\tilde{E}_i. \tag{7.4.7}$$

On the other hand, the field components for the longitudinal wave, i.e., the electron plasma wave are represented as

$$E_z = \frac{1}{2}\left[\tilde{E}_l e^{j(\omega_l t - k_l z)} + c.c.\right], \tag{7.4.8}$$

$$v_z = \frac{1}{2}\left[\tilde{v}_l e^{j(\omega_l t - k_l z)} + c.c.\right], \tag{7.4.9}$$

$$n = \frac{1}{2}\left[\tilde{n}_l e^{j(\omega_l t - k_l z)} + c.c.\right], \tag{7.4.10}$$

where $\tilde{E}_l, \tilde{v}_l$, and $\tilde{n}_l$ are assumed to vary slowly with z and t, satisfying the same condition as (7.4.6).

For the first-order quantities with respect to the electron plasma wave, we have the following amplitude relations:

$$\tilde{v}_l = j\frac{e}{\gamma_0^2 m_0 \omega_p}\tilde{E}_l, \quad \tilde{n}_l = j\frac{\varepsilon_0 k_l}{e}\tilde{E}_l. \tag{7.4.11}$$

Now we are ready to find the coupled-mode equations for the pump, scattered and electron plasma waves. First, substituting (7.4.3) ~ (7.4.5) and (7.4.8) ~ (7.4.10) in (7.3.5) and (7.3.2), and using the relations (7.4.7) and (7.4.11), we get for the complex amplitudes of the current densities $\tilde{J}_i$ and $\tilde{J}_s$ for the pump and scattered waves,

$$\tilde{J}_m = -j\frac{\varepsilon_0 \omega_p^2}{\omega_m}\tilde{E}_m + \varepsilon_0\left(\frac{\omega_p}{\omega_m}\right)^2 \frac{\partial \tilde{E}_m}{\partial t} + \tilde{J}_m^{NL}, \quad (m = i, s), \tag{7.4.12}$$

where

$$\tilde{J}_i^{NL} = -\frac{\varepsilon_0 \omega_p e}{2\gamma_0 m_0 \omega_s (\omega_i - v_0 k_i)} \left[\frac{k_s}{\gamma_0} + \beta_0 \frac{\omega_p}{c} - \gamma_0 \left(k_s - \beta_0 \frac{\omega_s}{c} \right) \right] \tilde{E}_s \tilde{E}_l + \frac{\varepsilon_0 e k_l}{2\gamma_0 m_0 \omega_s} \tilde{E}_s \tilde{E}_l, \tag{7.4.13}$$

$$\tilde{J}_s^{NL} = -\frac{\varepsilon_0 \omega_p e}{2\gamma_0 m_0 \omega_i (\omega_s - v_0 k_s)} \left[-\frac{k_i}{\gamma_0} + \beta_0 \frac{\omega_p}{c} + \gamma_0 \left(k_l - \beta_0 \frac{\omega_i}{c} \right) \right] \tilde{E}_i \tilde{E}_l^* - \frac{\varepsilon_0 e k_l}{2\gamma_0 m_0 \omega_i} \tilde{E}_i \tilde{E}_l^*, \tag{7.4.14}$$

$\tilde{J}_i^{NL}$ denoting the second-order current density for the pump wave produced by the coupling of the scattered and electron plasma waves, and $\tilde{J}_s^{NL}$ the second-order current density for the scattered wave produced by the coupling of the pump and electron plasma waves.

Using the current densities (7.4.12) in the second equation of (7.3.1), we obtain the following coupled mode equations:

$$\frac{\partial \tilde{E}_m}{\partial t} + v_{gm} \frac{\partial \tilde{E}_m}{\partial z} = -\frac{1}{2\varepsilon_0} \tilde{J}_m^{NL}, \quad (m = i, s), \tag{7.4.15}$$

with

$$v_{gm} = \frac{\partial \omega_m}{\partial k_m} = \frac{c^2 k_m}{\omega_m}, \quad (m = i, s), \tag{7.4.16}$$

v_{gm} $(m = i, s)$ being the group velocities of the pump and scattered waves.

Next, substitute the expressions (7.4.3) ~ (7.4.5) and (7.4.8) ~ (7.4.10) in (7.3.6) and (7.3.4), and use the relations (7.4.7) and (7.4.11) for the electromagnetic and electron plasma waves. Then, we obtain the complex amplitudes for the charge and current densities $\tilde{\rho}_l$ and $\tilde{J}_l$,

$$\tilde{\rho}_l = -j\varepsilon_0 k_l \tilde{E}_l + \frac{2\gamma_0 \varepsilon_0 k_l}{\omega_p} \frac{\partial \tilde{E}_l}{\partial t} + \varepsilon_0 \left(\frac{2\gamma_0 v_0 k_l}{\omega_p} + 1 \right) \frac{\partial \tilde{E}_l}{\partial z} + \tilde{\rho}_l^{NL}, \tag{7.4.17}$$

$$\tilde{J}_l = -j\varepsilon_0 \omega_l \tilde{E}_l + \varepsilon_0 \left(\frac{2\gamma_0 \omega_l}{\omega_p} - 1 \right) \frac{\partial \tilde{E}_l}{\partial t} + \frac{2\gamma_0 \varepsilon_0 \omega_l v_0}{\omega_p} \frac{\partial \tilde{E}_l}{\partial z} + \tilde{J}_l^{NL}, \tag{7.4.18}$$

$\tilde{\rho}_l^{NL}$ and $\tilde{J}_l^{NL}$ being the second-order charge and current densities for the electron plasma wave generated by the coupling of the pump and scattered waves, taking the forms

$$\tilde{\rho}_l^{NL} = \frac{\gamma_0 \varepsilon_0 e k_l}{2 m_0 \omega_i \omega_s} \left[\left(k_s - \beta_0 \frac{\omega_s}{c} \right) - \left(k_i - \beta_0 \frac{\omega_i}{c} \right) \right] \tilde{E}_i \tilde{E}_s^*, \tag{7.4.19}$$

$$\tilde{J}_l^{NL} = \frac{\omega_l}{k_l} \tilde{\rho}_l^{NL}. \tag{7.4.20}$$

Inserting $\tilde{\rho}_l$ in the second equation of (7.3.3), or $\tilde{J}_l$ in the first equation of (7.3.3), we find the coupled-mode equation for the electron plasma wave,

$$\frac{\partial \tilde{E}_l}{\partial t} + v_0 \frac{\partial \tilde{E}_l}{\partial z} = - \frac{\omega_p}{2 \gamma_0 \varepsilon_0 k_l} \tilde{\rho}_l^{NL}. \tag{7.4.21}$$

Equations (7.4.15) and (7.4.21) constitute a set of coupled-mode equations for the pump, scattered and electron plasma waves.

As a special case, assume that the amplitudes of the pump, scattered and electron plasma waves are temporally invariable. Then, we have from (7.4.15) and (7.4.21),

$$\frac{d\tilde{E}_i}{dz} = -K_i \tilde{E}_s \tilde{E}_l, \quad K_i > 0, \tag{7.4.22}$$

$$\frac{d\tilde{E}_s}{dz} = -K_s \tilde{E}_i \tilde{E}_l^*, \quad K_s > 0, \tag{7.4.23}$$

$$\frac{d\tilde{E}_l}{dz} = -K_l \tilde{E}_i \tilde{E}_s^*, \quad K_l > 0, \tag{7.4.24}$$

where

$$K_i = \frac{e}{4 \gamma_0 m_0 c^2} \left(\frac{\omega_i}{\omega_s} + \frac{\sqrt{\varepsilon_s}}{\sqrt{\varepsilon_i}} \right) + \frac{\beta_0 e \omega_p}{4 \gamma_0 m_0 c^2 \omega_i \sqrt{\varepsilon_i}} \frac{(\omega_p / \omega_s) + \gamma_0 \left(1 - \beta_0 \sqrt{\varepsilon_s}\right)}{1 + \beta_0 \sqrt{\varepsilon_i}}, \tag{7.4.25}$$

$$K_s = \frac{e}{4 \gamma_0 m_0 c^2} \left(\frac{\omega_s}{\omega_i} + \frac{\sqrt{\varepsilon_i}}{\sqrt{\varepsilon_s}} \right) - \frac{\beta_0 e \omega_p}{4 \gamma_0 m_0 c^2 \omega_s \sqrt{\varepsilon_s}} \frac{(\omega_p / \omega_i) - \gamma_0 \left(1 + \beta_0 \sqrt{\varepsilon_i}\right)}{1 - \beta_0 \sqrt{\varepsilon_s}}, \tag{7.4.26}$$

$$K_l = \frac{e \omega_p}{4 m_0 c^2 \beta_0} \left[\left(\frac{\sqrt{\varepsilon_s}}{\omega_i} + \frac{\sqrt{\varepsilon_i}}{\omega_s} \right) - \beta_0 \left(\frac{1}{\omega_i} - \frac{1}{\omega_s} \right) \right]. \tag{7.4.27}$$

If the mode amplitude of the pump wave is constant and $K_s K_l > 0$, the coupled-mode equations (7.4.23) and (7.4.24) have apparently solutions of exponentially growing waves. Then, the spatial growth rate for the scattered and electron plasma waves is given by

$$\Gamma = \sqrt{K_s K_l}\left|\tilde{E}_i(L)\right|, \tag{7.4.28}$$

where L denotes the interaction distance which must be short enough for the exponential growth of the scattered and electron plasma waves to be allowed.

7.5 Solutions of Coupled-Mode Equations

In the foregoing section, we have shown that the coupled-mode equations (7.4.23) and (7.4.24) have exponentially growing solutions for the constant-amplitude pump wave. In this section, we will try to find solutions for the coupled-mode equations (7.4.22) ~ (7.4.24) for the general case where the amplitude of the pump wave is allowed to vary along the longitudinal direction as the parametric interaction of three waves proceeds. For this purpose, let us first transform variables in the coupled-mode equations (7.4.22) ~ (7.4.24) according to

$$\frac{\tilde{E}_i(z)}{\left|\tilde{E}_i(L)\right|} = u_i e^{j\varphi_i}, \tag{7.5.1}$$

$$\sqrt{\frac{K_i}{K_s}}\frac{\tilde{E}_s(z)}{\left|\tilde{E}_i(L)\right|} = u_s e^{j\varphi_s}, \tag{7.5.2}$$

$$\sqrt{\frac{K_i}{K_l}}\frac{\tilde{E}_l(z)}{\left|\tilde{E}_i(L)\right|} = u_l e^{j\varphi_l}, \tag{7.5.3}$$

$$\zeta = \Gamma z, \tag{7.5.4}$$

where Γ is the spatial growth rate for the linear regime defined in (7.4.28). The variables u_m and φ_m ($m = i, s, l$) are assumed to take real values.

On applying the transformations for variables (7.5.1) ~ (7.5.4) to the coupled-mode equations (7.4.22) ~ (7.4.24), we get the following equations:

$$\frac{du_i}{d\zeta} = -u_s u_l \sin\theta, \tag{7.5.5}$$

$$\frac{du_s}{d\zeta} = -u_l u_i \sin\theta, \tag{7.5.6}$$

$$\frac{du_l}{d\zeta} = -u_i u_s \sin\theta, \tag{7.5.7}$$

$$\frac{d\theta}{d\zeta} = -\left(\frac{u_s u_l}{u_i} + \frac{u_l u_i}{u_s} + \frac{u_i u_s}{u_l}\right)\cos\theta, \tag{7.5.8}$$

$$\theta = \varphi_i - \varphi_s - \varphi_l + \frac{\pi}{2}. \tag{7.5.9}$$

Inserting (7.5.5) ~ (7.5.7) in (7.5.8), we get

$$\frac{d\theta}{d\zeta} = \left(\frac{1}{u_i}\frac{du_i}{d\zeta} + \frac{1}{u_s}\frac{du_s}{d\zeta} + \frac{1}{u_l}\frac{du_l}{d\zeta}\right)\cot\theta. \tag{7.5.10}$$

Upon integrating the above equation, we find

$$u_i u_s u_l \cos\theta = A, \tag{7.5.11}$$

where A is an integration constant.

From (7.5.5) ~ (7.5.7), we obtain the conservation equations

$$u_i^2 - u_s^2 = m_1 > 0, \tag{7.5.12}$$

$$u_i^2 - u_l^2 = m_2 > 0, \tag{7.5.13}$$

where m_1 and m_2 are integration constants. Substituting (7.5.11) ~ (7.5.13) into (7.5.5) yields

$$\frac{du_i^2}{d\zeta} = 2\sqrt{\left(u_i^2 - u_{i1}^2\right)\left(u_i^2 - u_{i2}^2\right)\left(u_i^2 - u_{i3}^2\right)}, \tag{7.5.14}$$

where u_{i1}^2, u_{i2}^2, and u_{i3}^2 are the three roots of

$$u_i^2\left(u_i^2 - m_1\right)\left(u_i^2 - m_2\right) - A^2 = 0, \tag{7.5.15}$$

with the condition $u_{i3}^2 \geq u_{i2}^2 \geq u_{i1}^2$. Since $u_i^2 > m_1$ from (7.5.12), $u_i^2 > m_2$ from (7.5.13), and the argument of the square root in (7.5.14) must be positive, we have $u_i^2 > u_{i3}^2$.

Now, if we change variables according to

$$\xi = \sqrt{\frac{u_{i3}^2 - u_{i1}^2}{u_i^2 - u_{i1}^2}}, \tag{7.5.16}$$

we have

$$0 < \xi < 1, \tag{7.5.17}$$

and Eq. (7.5.14) becomes

$$\frac{d\zeta}{d\xi} = -\frac{1}{\sqrt{u_{i3}^2 - u_{i1}^2}} \frac{1}{\sqrt{\left(1-\xi^2\right)\left(1-\kappa^2\xi^2\right)}}, \tag{7.5.18}$$

with

$$\kappa^2 = \frac{u_{i2}^2 - u_{i1}^2}{u_{i3}^2 - u_{i1}^2} \le 1. \tag{7.5.19}$$

Integrating (7.5.18) with respect to ξ, we get

$$\zeta - \zeta_L = -\frac{1}{\sqrt{u_{i3}^2 - u_{i1}^2}} \int_{\xi_L}^{\xi} \frac{d\xi}{\sqrt{\left(1-\xi^2\right)\left(1-\kappa^2\xi^2\right)}}, \tag{7.5.20}$$

where ζ_L and ξ_L are the values of ζ and ξ at $z = L$. From (7.5.20), we obtain

$$\xi^2 = sn^2\left[\sqrt{u_{i3}^2 - u_{i1}^2}\left(\zeta_L - \zeta + \alpha_L\right), \kappa^2\right], \tag{7.5.21}$$

where

$$\alpha_L = \frac{1}{\sqrt{u_{i3}^2 - u_{i1}^2}} \int_0^{\xi_L} \frac{d\xi}{\sqrt{\left(1-\xi^2\right)\left(1-\kappa^2\xi^2\right)}}, \tag{7.5.22}$$

and sn denotes the Jacobi elliptic function.

On applying the boundary conditions

$$u_s = u_s(0), \quad u_l = u_l(0) \qquad :z = 0, \tag{7.5.23}$$

$$u_i = u_i(L) = 1 \qquad :z = L, \tag{7.5.24}$$

the solutions for u_i^2, u_s^2, and u_l^2 can be expressed from (7.5.21), (7.5.16), (7.5.12), and (7.5.13) as

$$u_i^2 = 1 + \left(u_{i3}^2 - u_{i1}^2\right)\left(\frac{1}{\xi^2} - \frac{1}{\xi_L^2}\right), \tag{7.5.25}$$

$$u_s^2 = u_s^2(0) + \left(u_{i3}^2 - u_{i1}^2\right)\left(\frac{1}{\xi^2} - \frac{1}{\xi_0^2}\right), \tag{7.5.26}$$

$$u_l^2 = u_l^2(0) + \left(u_{i3}^2 - u_{i1}^2\right)\left(\frac{1}{\xi^2} - \frac{1}{\xi_0^2}\right), \tag{7.5.27}$$

where ξ_0 is the value of ξ at $z = 0$.

Referring to (7.5.20), we can obtain the relation between the interaction length L and the electric field strength of the pump wave at $z = L$ as

$$\zeta_L = \frac{1}{\sqrt{u_{i3}^2 - u_{i1}^2}} \int_{\xi_L}^{\xi_0} \frac{d\xi}{\sqrt{\left(1-\xi^2\right)\left(1-\kappa^2\xi^2\right)}} . \tag{7.5.28}$$

7.6 Energy Relations

In this section, we formulate the energy relations for the pump wave, the scattered wave and the electron plasma wave, discussing energy exchange among these waves. First, from (7.3.1), we find the following identity for the transverse waves (pump and scattered waves):

$$\frac{\partial}{\partial z}\left(\frac{1}{\mu_0} E_x B_y\right) + \frac{\partial}{\partial t}\left(\frac{1}{2}\varepsilon_0 E_x^2 + \frac{1}{2\mu_0} B_y^2\right) = -J_x E_x . \tag{7.6.1}$$

Equation (7.6.1) represents the conservation relation for instantaneous values of the transverse waves. On the other hand, it is the conservation relation for the physical quantities averaged over an appropriate time interval that is physically meaningful in dispersive media such as electron beams. Hence, averaging (7.6.1) over one period of the pump wave or the scattered wave, with the aid of (7.4.12), we get the energy conservation relations for the pump and scattered waves:

$$\frac{\partial \tilde{W}_m}{\partial t} + \frac{\partial \tilde{S}_m}{\partial z} = -\frac{1}{4}\left(\tilde{E}_m^* \tilde{J}_m^{NL} + c.c.\right), \tag{7.6.2}$$

$$\tilde{W}_m = \frac{1}{2}\varepsilon_0 \tilde{E}_m \tilde{E}_m^*, \tag{7.6.3}$$

$$\tilde{S}_m = \frac{1}{2}\frac{k_m}{\mu_0 \omega_m} \tilde{E}_m \tilde{E}_m^*, \tag{7.6.4}$$

$(m = i, s)$,

$\tilde{W}_m$ $(m = i, s)$ being the time-average energy densities for the pump and scattered waves, and $\tilde{S}_m$ $(m = i, s)$ the time-average power flow densities for them. The power flow densities $\tilde{S}_m$ are found to be related to the energy densities $\tilde{W}_m$ through

$$\tilde{S}_m = v_{gm} \tilde{W}_m . \tag{7.6.5}$$

In view of the relation (7.6.5), the energy conservation relation (7.6.2) can also be obtained directly from the coupled-mode equation (7.4.15).

For the electron plasma wave, on the other hand, we have from the first equation of (7.3.3),

$$\frac{\partial}{\partial t}\left(\frac{1}{2}\varepsilon_0 E_z^2\right) = -J_z E_z. \tag{7.6.6}$$

Averaging (7.6.6) over one period of the electron plasma wave, with the aid of (7.4.18), we obtain the energy conservation relation for the electron plasma wave,

$$\frac{\partial \tilde{W}_l}{\partial t} + \frac{\partial \tilde{S}_l}{\partial z} = -\frac{1}{4}\left(\tilde{E}_l^* \tilde{J}_l + c.c.\right), \tag{7.6.7}$$

$$\tilde{W}_l = -\frac{1}{2}\gamma_0 \frac{|\omega_l|}{\omega_p}\varepsilon_0 \tilde{E}_l \tilde{E}_l^*, \tag{7.6.8}$$

$$\tilde{S}_l = -v_0\left(\frac{1}{2}\gamma_0 \frac{|\omega_l|}{\omega_p}\varepsilon_0 \tilde{E}_l \tilde{E}_l^*\right), \tag{7.6.9}$$

$\tilde{W}_l$ being the time-average energy density for the electron plasma wave, and $\tilde{S}_l$ the time-average power flow density for it. We can also find the relation (7.6.7) directly from the coupled-mode equation (7.4.21). Note that $\tilde{W}_l$ and $\tilde{S}_l$ are negative for our analysis, since we are dealing with the negative-energy electron plasma wave.

It should be noted that Eqs. (7.6.2) and (7.6.7) represent the separate energy conservation relations for the pump, scattered and electron plasma waves. Let us now consider the overall energy conservation relation with all three waves included. For this purpose, we rewrite (7.6.2) and (7.6.7) as follows:

$$\frac{\partial \tilde{W}_i}{\partial t} + \frac{\partial \tilde{S}_i}{\partial z} = C_i\left(\tilde{E}_i \tilde{E}_s^* \tilde{E}_l^* + c.c.\right), \tag{7.6.10}$$

$$\frac{\partial \tilde{W}_s}{\partial t} + \frac{\partial \tilde{S}_s}{\partial z} = -C_s\left(\tilde{E}_i \tilde{E}_s^* \tilde{E}_l^* + c.c.\right), \tag{7.6.11}$$

$$\frac{\partial \tilde{W}_l}{\partial t} + \frac{\partial \tilde{S}_l}{\partial z} = -C_l\left(\tilde{E}_i \tilde{E}_s^* \tilde{E}_l^* + c.c.\right), \tag{7.6.12}$$

where

$$C_i = \frac{1}{2}c\varepsilon_0\sqrt{\varepsilon_i}K_i, \tag{7.6.13}$$

$$C_s = \frac{1}{2}c\varepsilon_0\sqrt{\varepsilon_s}K_s, \tag{7.6.14}$$

$$C_l = \frac{1}{2}\gamma_0 \varepsilon_0 v_0 \frac{\omega_l}{\omega_p} K_l. \tag{7.6.15}$$

From (7.6.13) ~ (7.6.15), we can prove that the following relation holds:

$$C_s + C_l = C_i. \tag{7.6.16}$$

Using the relation (7.6.16) in (7.6.10) ~ (7.6.12), we get the overall energy conservation relation with all the pump, scattered and electron plasma waves included:

$$\frac{\partial \tilde{W}}{\partial t} + \frac{\partial \tilde{S}}{\partial z} = 0, \tag{7.6.17}$$

$$\tilde{W} = \tilde{W}_i + \tilde{W}_s + \tilde{W}_l, \tag{7.6.18}$$

$$\tilde{S} = \tilde{S}_i + \tilde{S}_s + \tilde{S}_l. \tag{7.6.19}$$

From (7.6.13) ~ (7.6.15), we can also obtain the relations

$$\frac{C_s}{C_i} = \frac{\omega_s}{\omega_i}, \quad \frac{C_l}{C_i} = \frac{\omega_l}{\omega_i}, \tag{7.6.20}$$

which correspond to the Manley-Rowe relations for the Raman-regime free-electron laser.

In the foregoing discussion, we have derived the general conservation relations for the parametric interaction among the pump, scattered and electron plasma waves. However, for the free-electron laser, we usually treat a relativistic electron beam with $\beta_0 \cong 1$. Then, we have $|C_i| << |C_s|, |C_l|$, and $C_s \cong -C_l$. Specifically, for the case of $\beta_0 \cong 1$, we get from (7.6.13) ~ (7.6.15)

$$C_i \cong \frac{\varepsilon_0 e}{8\gamma_0 m_0 c}, \quad C_s \cong \frac{\varepsilon_0 e}{8\gamma_0 m_0 c}\frac{\omega_s}{\omega_i}, \quad C_l \cong \frac{\varepsilon_0 e}{8\gamma_0 m_0 c}\frac{\omega_l}{\omega_i}. \tag{7.6.21}$$

In deriving (7.6.21), we also assumed $\omega_p/\omega_i << 1$, for which $\varepsilon_i \cong 1$ and $\varepsilon_s \cong 1$. Hence, for the Raman-regime free-electron laser, we can consider that energy is exchanged almost exclusively between the scattered wave (positive-energy wave) and the electron plasma wave (negative-energy wave) and that the pump wave only plays the role of coupling these two waves.

For the special case, let us assume that the amplitudes of the pump, scattered and electron plasma waves are temporally invariable. Then, from (7.6.10) ~ (7.6.12), we have

$$\frac{d\tilde{S}_i}{dz} = C_i\left(\tilde{E}_i \tilde{E}_s^* \tilde{E}_l^* + c.c.\right), \tag{7.6.22}$$

$$\frac{d\tilde{S}_s}{dz} = -C_s\left(\tilde{E}_i\tilde{E}_s^*\tilde{E}_l^* + c.c.\right), \tag{7.6.23}$$

$$\frac{d\tilde{S}_l}{dz} = -C_l\left(\tilde{E}_i\tilde{E}_s^*\tilde{E}_l^* + c.c.\right). \tag{7.6.24}$$

The solutions for the scattered and electron plasma waves obtained from (7.6.22) ~ (7.6.24) can be expressed as

$$\tilde{S}_s(L) - \tilde{S}_s(0) = -\frac{C_s}{C_i}\left[\tilde{S}_i(L) - \tilde{S}_i(0)\right], \tag{7.6.25}$$

$$\tilde{S}_l(L) - \tilde{S}_l(0) = \frac{C_l}{C_i}\left[\tilde{S}_i(L) - \tilde{S}_i(0)\right]. \tag{7.6.26}$$

When we treat laser oscillation, we can generally assume that the following conditions hold:

$$\left|\tilde{S}_i(0)\right| << \left|\tilde{S}_i(L)\right|, \ \tilde{S}_s(0) << \tilde{S}_s(L), \ \left|\tilde{S}_l(0)\right| << \left|\tilde{S}_l(L)\right|. \tag{7.6.27}$$

Then, the maximum power flow density for the scattered wave can be obtained by applying the above conditions to (7.6.25) as

$$\max\left\{\tilde{S}_s(L)\right\} = \frac{C_s}{C_i}\left|\tilde{S}_i(L)\right|. \tag{7.6.28}$$

The value of ζ_L to get the maximum power flow density for the scattered wave can be found from (7.5.28),

$$\max\left\{\zeta_L\right\} = \frac{1}{\sqrt{u_{i3}^2 - u_{i1}^2}}\int_0^{\xi_0} \frac{d\xi}{\sqrt{\left(1-\xi^2\right)\left(1-\kappa^2\xi^2\right)}}. \tag{7.6.29}$$

Let us discuss quantitatively how the power flow densities for the pump, scattered and electron plasma waves vary as they propagate along the electron beam. For this purpose, with the aid of (7.6.4) and (7.6.9), together with (7.5.1) ~ (7.5.3), we rewrite their power flow densities as follows:

$$\tilde{S}_i = -\left|\tilde{S}_i(L)\right|u_i^2, \tag{7.6.30}$$

$$\tilde{S}_s = \frac{C_s}{C_i}\left|\tilde{S}_i(L)\right|u_s^2 = \frac{\omega_s}{\omega_i}\left|\tilde{S}_i(L)\right|u_s^2, \tag{7.6.31}$$

$$\tilde{S}_l = \frac{C_l}{C_i}\left|\tilde{S}_i(L)\right|u_l^2 = -\frac{\left|\omega_l\right|}{\omega_i}\left|\tilde{S}_i(L)\right|u_l^2 \cong -\frac{\left|\omega_l\right|}{\omega_i}\left|\tilde{S}_i(L)\right|u_s^2. \tag{7.6.32}$$

For the case of $\beta_0 = 0.98$ and $\omega_p/\omega_i = 0.2$, we show in Fig.7.4 the dependence of the power flow densities upon the interaction length. In the above numerical example, we have for the values of ω_s/ω_i and ζ_L,

$$\omega_s/\omega_i = 96, \quad \zeta_L = 5.17. \tag{7.6.33}$$

The value of $\zeta_L = 5.17$ corresponds to

$$\left|\tilde{E}_i(L)\right|L = 2.64 \times 10^7 \text{V}. \tag{7.6.34}$$

Assuming $\omega_i/2\pi$ = 10GHz in (7.6.33), we get $\omega_s/2\pi$ = 960GHz (wavelength 0.31 mm). In addition, let the magnitude of the pump field be 8.8×10^6 V/m in (7.6.34). Then, the interaction length becomes L = 3m.

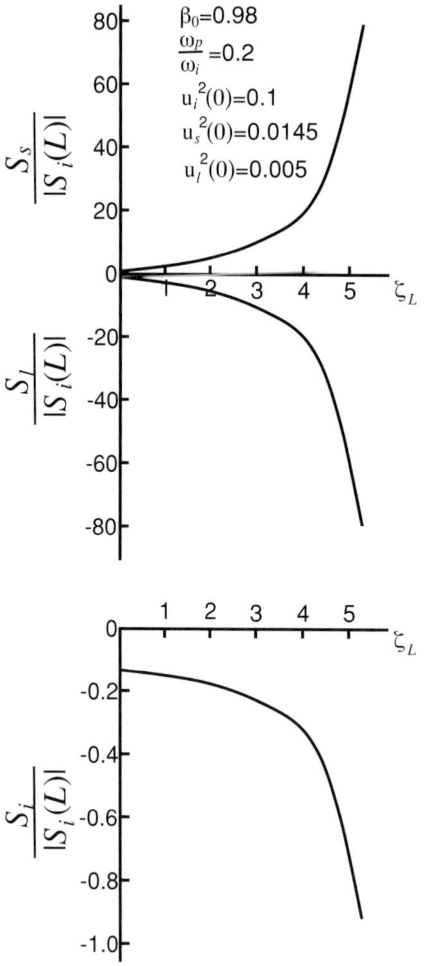

Fig. 7.4. Power flow densities of pump, scattered and electron plasma waves

7.7 Saturation in Laser Output and Efficiency of Energy Transfer

The output of the free-electron laser is suppressed by the saturation mechanisms. The first mechanism for saturation in the output power is one that is based upon the depletion of the pump wave. Specifically, due to the coupling of the scattered and electron plasma waves, a nonlinear current density with the same frequency and wave number as the pump wave is produced. It causes, in turn, the amplitude of the pump wave to be damped. The other saturation mechanism is one that results from trapping of electrons in the electron plasma wave. Since the former saturation mechanism has been discussed in detail in the preceding sections, we now investigate the latter saturation mechanism and the efficiency of energy transfer from the kinetic energy of the electron beam to the electromagnetic wave energy based upon it. In order to realize a free-electron laser with high power and high efficiency, in particular, the latter saturation mechanism becomes an important factor to be considered.

The saturation mechanism due to the trapping of electrons in the electron plasma wave has already been well known for microwave tubes such as the traveling wave tube. We have also discussed in detail bunching and trapping of electrons in electric field in connection with the saturation in electromagnetic wave growth in Chaps. 5 and 6. Therefore, we describe only briefly its essence in the following discussion, leaving a detailed treatment to elsewhere [7.13]. As seen from Fig. 7.3, the phase velocity of the electron plasma wave (negative-energy wave) is less than the drift velocity of the electron beam. Hence electrons are bunched in the decelerating phase of the electron plasma wave, and energy is transferred from the bunched electrons to the electromagnetic scattered wave, as a result of which the latter grows. The bunched electrons grow gradually under the influence of the pump wave and the scattered wave. However, at the same time, they are gradually slowed down as they give up energy to the scattered wave until they finally cease to deliver energy to it. The laser output gets saturated at this limiting velocity of the bunched electrons. Let the phase velocity of the electron plasma wave be denoted by v_l. Then, the average velocity of the electron beam v_{0s} at which the laser output becomes saturated is given by [7.13]

$$v_0 - v_{0s} = 2\left(v_0 - v_l\right), \tag{7.7.1}$$

where

$$v_l = v_0 - c\frac{1}{\gamma_0}\frac{\omega_p}{\omega_i\sqrt{\varepsilon_i} + \omega_s\sqrt{\varepsilon_s}}, \tag{7.7.2}$$

v_l being found with the aid of the dispersion relation for the electron plasma wave (7.4.2) and the phase-matching condition (7.4.1).

In the following discussion, we consider the efficiency of energy transfer from the kinetic energy of the electron beam to the laser output, assuming for simplicity that all the electrons constituting the electron beam have been slowed down at saturation to the velocity v_{0s} given by (7.7.1). First, the initial kinetic energy per unit volume of the electron beam W_0 is given by

$$W_0 = n_0 m_0 c^2 (\gamma_0 - 1). \tag{7.7.3}$$

At saturation, on the other hand, the kinetic energy per unit volume of the electron beam W_s reduces to

$$W_s = n_0 m_0 c^2 (\gamma_s - 1), \tag{7.7.4}$$

with

$$\gamma_s = \frac{1}{\sqrt{1-\beta_s^2}}, \quad \beta_s = \frac{v_{0s}}{c}. \tag{7.7.5}$$

Hence the kinetic energy corresponding to the difference $\Delta W = W_0 - W_s$ is converted to the electromagnetic wave energy. From the above discussion, the efficiency of energy transfer from the kinetic energy of the electron beam to the laser output, i.e., the electromagnetic wave energy η can be defined as

$$\eta = \frac{\Delta W}{W_0} = \frac{\gamma_0 - \gamma_s}{\gamma_0 - 1}. \tag{7.7.6}$$

Since $v_0 - v_{0s} << v_0$ in general, Eq. (7.7.6) can be rewritten as

$$\eta = \frac{2\gamma_0^3 \beta_0 (\beta_0 - \beta_l)}{\gamma_0 - 1}, \quad \beta_l = \frac{v_l}{c}. \tag{7.7.7}$$

For the special case of $\beta_0 \cong 1$ and $\omega_p << \omega_i$, η is simplified, with the aid of (7.7.2), to

$$\eta = \frac{1}{2(\gamma_0 - 1)} \frac{\omega_p}{\omega_i}. \tag{7.7.8}$$

For the case of $\beta_0 = 0.98$ and $\omega_p/\omega_i = 0.2$, we have $\eta = 2.5\%$.

8. FDTD Analysis of Beam-Wave Interaction

8.1 Introduction

In the theoretical treatment of beam-wave interactions, the fluid model or the single-particle model has been used for representing the relativistic electron beam, as described in the foregoing chapters. However, it is very difficult to clarify the details of complicated beam-wave interactions, which are generally nonlinear in essence, only with the aid of purely theoretical approaches based upon either of these models. In fact, with the rapid development of digital computers in recent years, computer simulation has become a powerful tool for the research of nonlinear physical phenomena in various electron beam devices. In this chapter, we describe a typical branch of computer simulation, what is called particle simulation [8.1–8.3], which treats resonant interactions between a collection of charged particles and electromagnetic waves in plasmas or electron beams. In this method, the Maxwell equations and the relativistic equation of motion are discretized in space and time for numerical analysis of electromagnetic wave propagation and particle motion. Thus the interaction of a collection of charged particles and electromagnetic waves is temporally followed with the aid of the finite-difference time-domain (FDTD) method [8.4, 8.5].

The validity of particle simulation in the nonlinear analysis of beam-wave interactions has been well confirmed by comparison with a large number of experimental results. Hence the method of particle simulation has become an indispensable tool for the analysis and design of various electron beam devices in recent years. In this chapter, we apply the method of particle simulation for the analysis and design of a Cherenkov laser as a specific example, in order to explain how to use this method.

8.2 Basic Equations for Particle Simulation

As a specific example of relativistic electron beam devices for numerical analysis of beam-wave interaction via particle simulation, let us consider a two-dimensional model of the Cherenkov laser shown in Fig. 8.1, together with the coordinate system. The 2-D model of the Cherenkov laser under consideration is composed of a

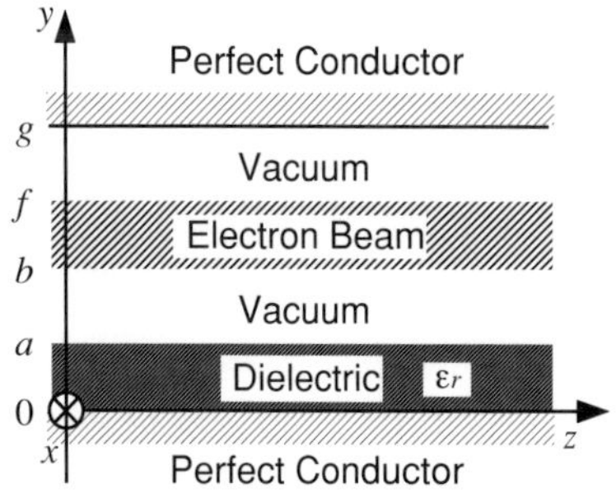

Fig. 8.1. 2-D model of the Cherenkov laser

parallel plate waveguide, one conducting plate of which is coated with a dielectric sheet with thickness a and relative permittivity ε_r, and a planar relativistic electron beam drifting in the waveguide with the average velocity $\langle v_{zi} \rangle$. In Fig. 8.1, the separation between two conducting plates is g, the thickness of a planar relativistic electron beam $f - b$, and the beam-dielectric gap $b - a$. For simplicity, the electron beam is confined in the direction of beam flow or in the z direction by an infinite static magnetic field. In addition, let the electron beam be neutralized by a flow of ions. Specifically, in order to cancel out the static electric and magnetic fields produced by electrons, we assume that ions with the same charge density as electrons are drifting in the background of the electron beam with the same velocity as the initial velocity of electrons v_0. Then, both electron and ion flows develop the Cherenkov instability. However, the Cherenkov instability due to the ion flow can be neglected as compared with that due to the electron flow. This is because the growth rate for the former is much less than for the latter (note that the growth rate for the Cherenkov instability is inversely proportional to the third power of the mass of charged particles).

The basic equations for particle simulation of beam-wave interaction in relativistic electron beam devices are the Maxwell equations and the relativistic equation of motion for the electron. For the case where the electron beam is immersed in an infinite magnetostatic field, it is not coupled with TE waves. Hence we consider only the propagation of TM waves in the following discussion. For TM waves propagated in a 2-D system as shown in Fig. 8.1, the Maxwell equations take the form

$$\frac{\partial E_z}{\partial y} - \frac{\partial E_y}{\partial z} = -\frac{\partial B_x}{\partial t}, \tag{8.2.1}$$

$$\frac{\partial B_x}{\partial z} = \frac{\varepsilon_r}{c^2}\frac{\partial E_y}{\partial t}, \tag{8.2.2}$$

$$-\frac{\partial B_x}{\partial y} = \frac{\varepsilon_r}{c^2}\frac{\partial E_z}{\partial t} + \mu_0 J_z, \tag{8.2.3}$$

where

$$J_z = J_z(y,z,t) = \sum_i q_i v_{zi} \delta(y - y_i) \delta(z - z_i), \tag{8.2.4}$$

c denoting the speed of light in vacuum, μ_0 the permeability of vacuum, and ε_r the relative permittivity of each region in Fig. 8.1. The quantity J_z represents the current density produced by two kinds of superparticles, one for electrons and the other for ions. In particle simulation, we introduce the concept of superparticles with large mass and charge, which are composed of a large number of actual particles. By following the motion of superparticles in an electromagnetic field, we can properly describe the collective behavior of actual particles in the field while greatly reducing the apparent number of particles involved and thus the time required for numerical simulation. We assume that superparticles for ions are a uniformly moving background for neutralizing the mean charge and current densities produced by electrons. In addition, q_i denotes the charge of the ith superparticle composed of a large number of electrons or ions, v_{zi} the velocity of the ith superparticle at the position (y_i, z_i) and at the time t, and δ the Dirac delta function. Note that we can get the field equations in each region of Fig. 8.1 from (8.2.1) ~ (8.2.3) by the following substitutes. Namely, setting $\varepsilon_r = 1$ in the above equations, we obtain the field equations in the region of the electron beam. Equations (8.2.1) ~ (8.2.3) with $\varepsilon_r = 1$ and $J_z = 0$ give the field equations in the vacuum regions, while putting $J_z = 0$ yields those in the dielectric region.

For the case where the electron beam is magnetically confined in the z direction by an infinite magnetostatic field, electrons constituting the electron beam can move only in the z direction. Then, the relativistic equation of motion for a superparticle for electrons is expressed as

$$\frac{d}{dt}\left(m_e u_{zi}\right) = q_e E_z\left(y_i, z_i, t\right), \tag{8.2.5}$$

with

$$u_{zi} = \gamma_i v_{zi}, \quad v_{zi} = \frac{dz_i}{dt}, \quad \gamma_i = \frac{1}{\sqrt{1 - \left(\frac{v_{zi}}{c}\right)^2}}, \tag{8.2.6}$$

where m_e and q_e denote the rest mass and the electric charge of a superparticle for electrons, respectively, which are integral multiples of the rest mass and the charge of the electron, m_0 and $-e$.

8.3 Particle Simulation

For numerical analysis of the interaction between an electromagnetic wave and the electron beam in the Cherenkov laser, we divide the 2-D model of the Cherenkov laser shown in Fig. 8.1 into many small segments (see Fig. 8.2 (a)), each of which is assumed to be one guide wavelength long ($L = \lambda_g$), in the longitudinal direction.

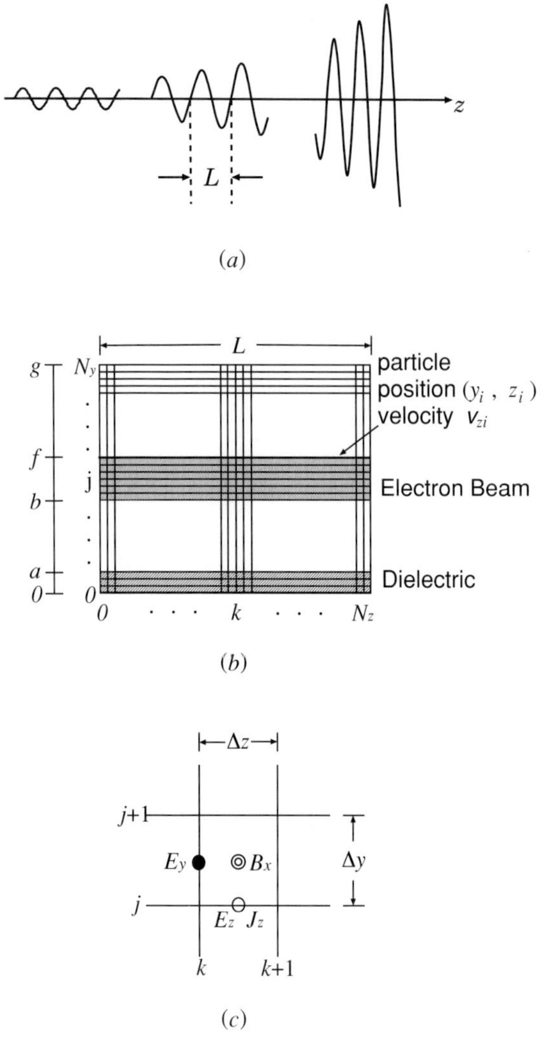

Fig. 8.2. FDTD analysis of beam-wave interaction. (*a*) Division of the entire system into small segments. (*b*) Subdivision of a small segment. (*c*) Discretization of field components

The guide wavelength λ_g is defined as the wavelength measured along the waveguide. We assume that the periodic boundary conditions are approximately satisfied between adjacent segments. The periodic boundary approximation is valid for the case where the growth rate for growing waves is sufficiently small, namely, it is so small that beam-wave interactions in adjacent segments proceed in almost the same way with almost the same amplitudes. Then, we pick out a group of particles contained in one particular segment around the origin at the initial state, and temporally follow, with the aid of the FDTD method, the interaction between an electromagnetic wave and the particular group of particles, as it travels down the waveguide. In order to discretize the Maxwell equations, as shown in Fig. 8.2 (*b*), each segment of the 2-D model for the Cherenkov laser is further subdivided into N_y subsegments in the y direction and N_z subsegments in the z direction. The spatial grids subdividing a small segment in the y direction are spaced by Δy and numbered by $j\,(j = 0,1,2,\cdots N_y)$. On the other hand, those in the z direction are spaced by Δz and numbered by $k\,(k = 1,2,3,\cdots N_z)$. We assume that the surfaces of the dielectric and the electron beam coincide with one of the spatial grids in the y direction. In the beam region of one particular segment, we uniformly arrange, at the initial state, two kinds of superparticles, one for electrons and the other for ions. Then, we follow the interaction between superparticles for electrons and the electromagnetic wave while successively increasing time t from $t = 0$ by an increment Δt. The time increment Δt must satisfy the stability condition for field calculation, what is called the Courant condition

$$(c\Delta t)^2\left[\frac{1}{(\Delta y)^2}+\frac{1}{(\Delta z)^2}\right]<1. \tag{8.3.1}$$

In order to represent the Maxwell equations in terms of centered spatial differencing, we choose the relative spatial locations of the TM field components as illustrated in Fig. 8.2 (*c*). Then, the finite difference time domain (FDTD) representations for the Maxwell equations are expressed as

$$\frac{B_x^{n+1/2}\left(j+\frac{1}{2},k+\frac{1}{2}\right)-B_x^{n-1/2}\left(j+\frac{1}{2},k+\frac{1}{2}\right)}{\Delta t}$$

$$=-\left[\frac{E_z^n\left(j+1,k+\frac{1}{2}\right)-E_z^n\left(j,k+\frac{1}{2}\right)}{\Delta y}-\frac{E_y^n\left(j+\frac{1}{2},k+1\right)-E_y^n\left(j+\frac{1}{2},k\right)}{\Delta z}\right], \tag{8.3.2}$$

$$\frac{E_y^{n+1}\left(j+\frac{1}{2},k\right)-E_y^n\left(j+\frac{1}{2},k\right)}{\Delta t}$$

$$=\frac{c^2}{\varepsilon_r}\frac{B_x^{n+1/2}\left(j+\frac{1}{2},k+\frac{1}{2}\right)-B_x^{n+1/2}\left(j+\frac{1}{2},k-\frac{1}{2}\right)}{\Delta z}, \tag{8.3.3}$$

$$\frac{E_z^{n+1}\left(j,k+\frac{1}{2}\right)-E_z^{n}\left(j,k+\frac{1}{2}\right)}{\Delta t}$$

$$=-\frac{c^2}{\varepsilon_r}\frac{B_x^{n+1/2}\left(j+\frac{1}{2},k+\frac{1}{2}\right)-B_x^{n+1/2}\left(j-\frac{1}{2},k+\frac{1}{2}\right)}{\Delta y}$$

$$-\frac{1}{\varepsilon_0\varepsilon_r}J_z^{n+1/2}\left(j,k+\frac{1}{2}\right), \tag{8.3.4}$$

where the abbreviation such as

$$B_x(j\Delta y,k\Delta z,n\Delta t)=B_x^n(j,k), \tag{8.3.5}$$

is used. The notation $B_x^n(j,k)$ denotes the x component of the magnetic flux density at the position ($j\Delta y$, $k\Delta z$) and at the time $n\Delta t$.

The current densities on the spatial grids are interpolated from the positions of superparticles as

$$J_{zl}^{n+1/2}=\sum_i \frac{1}{2}q_e\frac{u_{zi}^{n+1/2}}{\gamma_i}\left[S\left(\boldsymbol{r}_l-\boldsymbol{r}_i^{n+1}\right)+S\left(\boldsymbol{r}_l-\boldsymbol{r}_i^{n}\right)\right], \tag{8.3.6}$$

with

$$S\left(\boldsymbol{r}_l-\boldsymbol{r}_i^{m}\right)=S\left(y_l-y_i^{m}\right)S\left(z_l-z_i^{m}\right),(m=n,\,n{+}1). \tag{8.3.7}$$

The function $S(\boldsymbol{r}_l-\boldsymbol{r}_i)$ represents a shape factor for which we assume the quadratic spline, $\boldsymbol{r}_l$ and $\boldsymbol{r}_i$ being the position vectors for the position on spatial grids ($j\Delta y$, $k\Delta z$) and the position for a superparticle (y_i, z_i). The specific functional form for the quadratic spline is expressed as

$$S(X_l-X_i)=\frac{1}{\Delta X}\left[\frac{3}{4}-|\delta X|^2\right],\quad |\delta X|\le\frac{1}{2}, \tag{8.3.8}$$

$$S(X_l-X_i)=\frac{1}{2\Delta X}\left(\frac{3}{2}-|\delta X|\right)^2,\quad \frac{1}{2}<|\delta X|\le\frac{3}{2}, \tag{8.3.9}$$

$$S(X_l-X_i)=0,\quad |\delta X|>\frac{3}{2}, \tag{8.3.10}$$

with

$$\delta X=\frac{X_l-X_i}{\Delta X},\quad X=y,z. \tag{8.3.11}$$

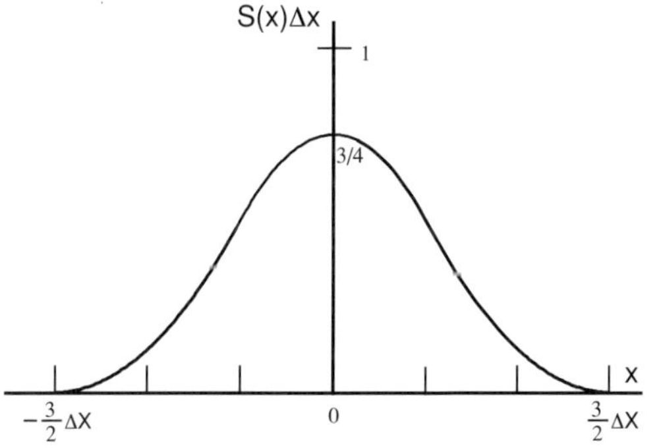

Fig. 8.3. Shape factor of the quadratic spline

The shape factor given by (8.3.8) ~ (8.3.10) is illustrated in Fig. 8.3. The shape factor S is introduced to replace a point source by a continuous spatial distribution, thus avoiding some numerical difficulties. Specifically, by replacing a point source by a continuous spatial distribution, we can suppress fluctuations or noises in charge density and fields, having more smoothed simulation results.

In a similar manner, the FDTD representation for the relativistic equation of motion for a superparticle for electrons at the position (y_i, z_i) is given by

$$\frac{u_{zi}^{n+1/2} - u_{zi}^{n-1/2}}{\Delta t} = \frac{q_e}{m_e} E_{zi}^{n}, \tag{8.3.12}$$

$$\frac{z_i^{n+1} - z_i^{n}}{\Delta t} = \frac{u_{zi}^{n+1.2}}{\gamma_i^{n+1/2}}, \tag{8.3.13}$$

where E_{zi} denotes the electric field component at the position of a superparticle, which can be interpolated from adjacent spatial grids as

$$E_{zi} = \sum_{l} E_{zl} S(\boldsymbol{r}_l - \boldsymbol{r}_i) \Delta y \Delta z\,. \tag{8.3.14}$$

Next, let us consider the boundary condition at the dielectric surface $y = j_1 \Delta y$. For this purpose, we apply the integral form of Ampere-Maxwell's law to a small rectangular contour C perpendicular to the boundary between the dielectric and vacuum regions and an open surface S bounded by the closed contour C, as shown in Fig. 8.4. Then, we have

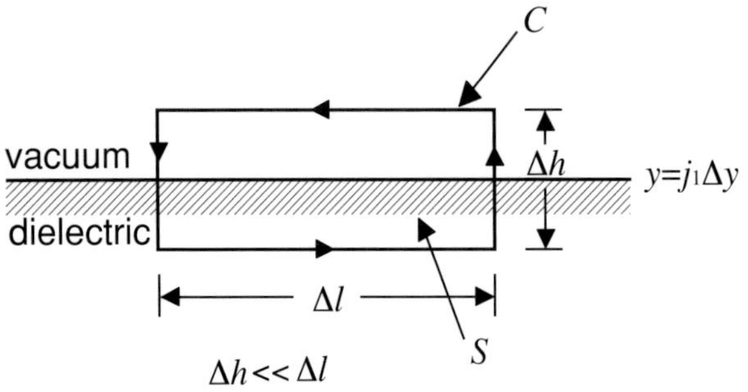

Fig. 8.4. Boundary condition at the dielectric surface

$$\frac{1}{\mu_0}\left[B_x\left(j_1-\frac{1}{2},k+\frac{1}{2}\right)-B_x\left(j_1+\frac{1}{2},k+\frac{1}{2}\right)\right]\Delta l$$

$$=\frac{d}{dt}\left[\varepsilon_0(1+\varepsilon_r)E_z\left(j_1,k+\frac{1}{2}\right)\Delta l\frac{\Delta y}{2}\right], \tag{8.3.15}$$

from which we get

$$\frac{B_x\left(j_1-\frac{1}{2},k+\frac{1}{2}\right)-B_x\left(j_1+\frac{1}{2},k+\frac{1}{2}\right)}{\Delta y}$$

$$=\frac{\varepsilon_r+1}{2c^2}\frac{d}{dt}\left[E_z\left(j_1,k+\frac{1}{2}\right)\right]. \tag{8.3.16}$$

The temporal differencing of (8.3.16) yields the following result which replaces (8.3.4) at the boundary between the dielectric and vacuum regions:

$$\frac{E_z^{n+1}\left(j_1,k+\frac{1}{2}\right)-E_z^{n}\left(j_1,k+\frac{1}{2}\right)}{\Delta t}$$

$$=-\frac{2c^2}{\varepsilon_r+1}\frac{B_x^{n+1/2}\left(j_1+\frac{1}{2},k+\frac{1}{2}\right)-B_x^{n+1/2}\left(j_1-\frac{1}{2},k+\frac{1}{2}\right)}{\Delta y}. \tag{8.3.17}$$

On the surfaces of conducting plates, on the other hand, the tangential component of electric field must vanish. Hence we have

$$E_z^n(0,k) = E_z^n(N_y,k) = 0, \tag{8.3.18}$$

for all values of n and k.

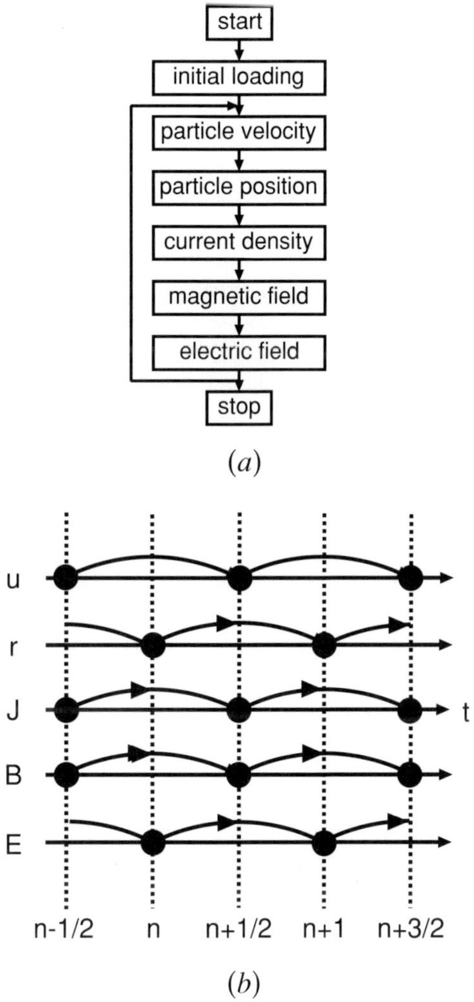

Fig. 8.5. Procedure for the FDTD calculation. (*a*) Flow chart for the procedure. (*b*) Time steps for calculating various physical quantities

If we are given the initial conditions on the field distributions for an electromagnetic wave, and those on the positions and velocities for superparticles, we can follow step by step the interaction between an electromagnetic wave and a group of superparticles with the aid of (8.3.2) ~ (8.3.4), and (8.3.12) and (8.3.13), together with (8.3.17) and (8.3.18), starting from $t = 0$ or $n = 0$. In Fig. 8.5, we show the flow chart of the procedure for numerical simulation and how to choose time steps for calculating various physical quantities. In this section, we have introduced a particle simulation scheme based upon the periodic boundary approximation. The advantage of this simulation scheme is that we can get useful simulation results in this scheme by a small amount of computational time and memory. For a more general case where the periodic boundary condition is not available, we must rely upon a simulation scheme which covers the entire region of beam-wave interaction. However, it should be noted that a much more amount of computational time and memory is required for the latter scheme than for the former scheme [8.11].

8.4 Nonlinear Beam-Wave Interaction in a Cherenkov Laser

Let us investigate the nonlinear characteristics of a 2-D model for the Cherenkov laser as shown in Fig. 8.1, with the aid of the particle simulation scheme developed in the foregoing section. The values of various parameters used in our numerical simulation are given in Table 8.1. For the value of the guide wavelength λ_g and the initial value of the electron drift velocity normalized by the velocity of light in vacuum, β_0, we choose those values in such a way that the velocity matching condition between the electromagnetic wave and the electron beam is satisfied and the growth rate in the linear analysis becomes maximum. In addition, at the initial state, we assume that an electromagnetic wave of the lowest-order TM mode with small enough amplitude exists and superparticles for electrons in a selected small segment are slightly modulated sinusoidally in terms of their positions and velocities.

In Fig. 8.6, we show how the amplitude of longitudinal electric field $|E_z|$ for the growing wave varies as the beam-wave interaction progresses. The value of $|E_z|$ is evaluated near the dielectric surface. As seen from Fig. 8.6, the electric field amplitude grows exponentially in the initial stage of the growth where the linear approximation holds. Then, the growth of the field amplitude is gradually suppressed and finally reaches saturation as the more particles get trapped in the longitudinal electric field and the nonlinear effect of the electron beam becomes predominant. The saturation time when the growth of the electromagnetic wave reaches saturation depends on the initial amplitude of the electric field. For the specific example in Fig. 8.6, the saturation time is equal to 1.699/c[s], which corresponds to the interaction length of 1.4 m. For comparison, we have also shown in Fig. 8.6 the numerical result obtained from the linear theoretical analysis.

Table 8.1. Values of various parameters used in Sect. 8.4

Relative Permittivity of Dielectric ε_r	2.12
Thickness of Dielectric a	0.5 [mm]
Beam Thickness (f - b)	0.125 [mm]
Beam-Dielectric Gap (b - a)	0.5 [mm]
Separation between Parallel Plates g	2.0 [mm]
Initial Beam Velocity Normalized by the Speed of Light β_0	0.872
Plasma Frequency for Electrons $\omega_p a/c$	0.01
Length of Beam Segment L (One Guide Wavelength λ_g)	2.0 [mm]
Number of Superparticles for Electrons N	256
Number of Spatial Grids in the y Direction N_y	64
Number of Spatial Grids in the z Direction N_z	64
Spatial Grid Interval $\Delta y = \Delta z$	0.03125 [mm]
Time Interval for One Step $c\Delta t$	0.000022 [mm]

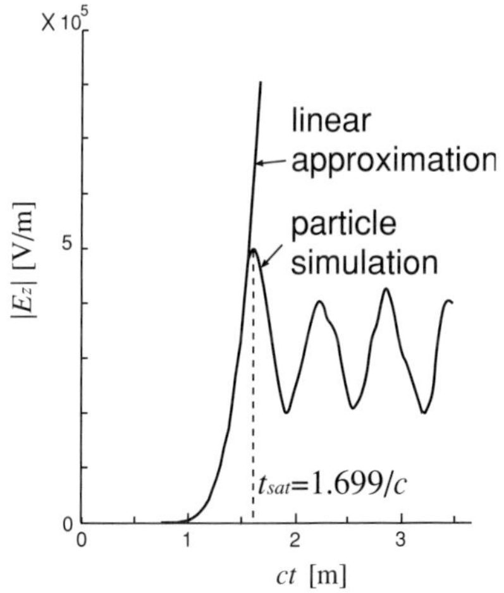

Fig. 8.6. Growth of the longitudinal electric field

Table 8.2. Comparison of particle simulation and linear analysis

	Particle Simulation	Linear Analysis
Normalized Frequency $\omega a/c$	1.29719	1.29724
Normalized Growth Rate $\alpha a/c$	0.00361	0.00364

In addition, in order to check the accuracy of the particle simulation, we compare in Table 8.2 the numerical results for the operating frequency and the temporal growth rate α obtained from the particle simulation and the linear theoretical analysis. This comparison is necessarily made in the region where the linear approximation is valid. From Table 8.2, we find that we can get sufficient accuracy with the method of particle simulation.

As stated in the above discussion, the nonlinearity appearing in the beam-wave interaction comes from trapping of electrons in the longitudinal electric field of the growing wave. As electrons get trapped in the electric field of the growing wave, the response of the electron beam to the electromagnetic wave comes to depend upon the wave amplitude. Specifically, the growth rate becomes dependent upon the amplitude of electric field as the nonlinear beam-wave interaction develops. Here, the growth rate is defined as the temporal or spatial derivative of the exponent for the growing wave represented in the form of exponential function. In Fig. 8.7, we illustrate how electrons are trapped and bunched in the phase space. In Fig. 8.7, the positions of particles are represented relative to the reference framework translating in the z direction with the velocity equal to the initial beam velocity v_0. In addition, the longitudinal length of the beam segment L is chosen equal to one guide wavelength of the electromagnetic wave. We initially assume that the electric field of the electromagnetic wave has an accelerating phase for electrons in the left half of the beam segment, and a decelerating phase in the right half of the beam segment. As seen from Fig. 8.7, a group of particles distributed uniformly in the beam segment at $t = 0$ gradually gets bunched in the decelerating phase of the electric field as time passes until it finally reaches saturation, and thereafter enters the accelerating phase of the electric field. When most of electrons are trapped in the decelerating phase of the electric field and the kinetic energy of a group of electrons becomes minimum, the field amplitude becomes maximum and saturated. As the beam-wave interaction proceeds further, energy is transferred back and forth between the electron beam and the electromagnetic wave. In Figs. 8.8 and 8.9 are shown the velocity distribution of particles and the variation of the average velocity of particles. Comparing the results in Figs. 8.6, 8.8 and 8.9, we see that the average velocity of a group of electrons decreases as it loses energy to the electromagnetic wave. From these figures, we can realize more clearly how the energy exchange between the electron beam and the electromagnetic wave occurs in the process of beam-wave interaction.

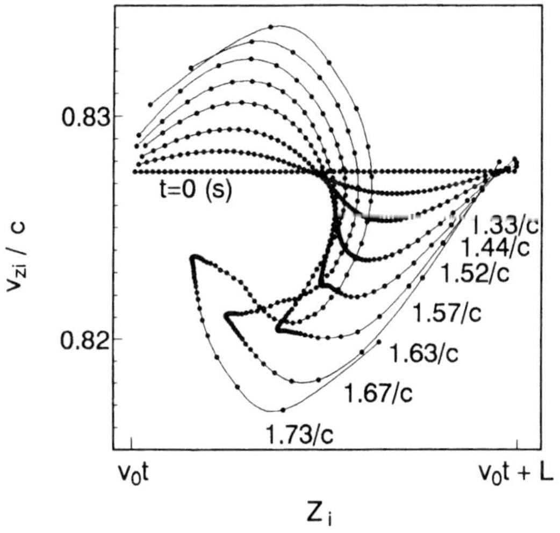

Fig. 8.7. Phase space diagram for trapping of particles in the electric field

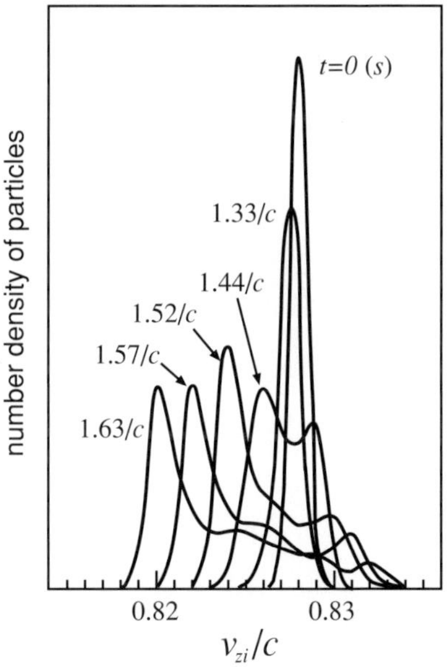

Fig. 8.8. Velocity distribution for particles

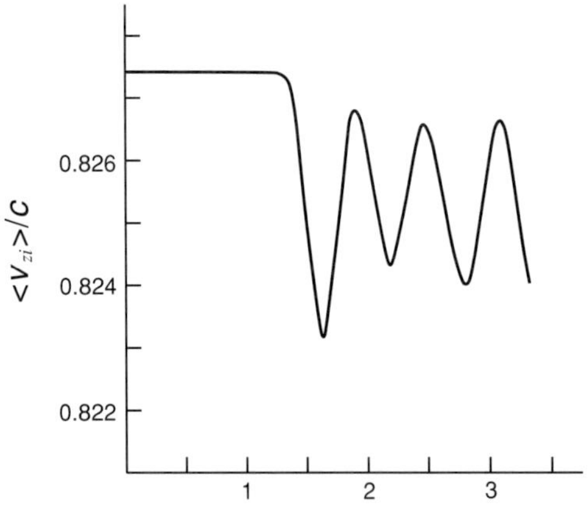

Fig. 8.9. Variation of the average velocity for par-ticles

Finally, we discuss the efficiency of energy transfer from the electron beam to the electromagnetic wave with the aid of numerical examples. First, at the saturation time, the total energy for the electromagnetic wave in a small segment per unit length in the x direction is given by

$$W = \sum_{j,k}\Big[\frac{1}{2}\varepsilon_0\varepsilon_r\left|E_y(j+\frac{1}{2},k)\right|^2 + \frac{1}{2}\varepsilon_0\varepsilon_r\left|E_z(j,k+\frac{1}{2})\right|^2$$

$$+\frac{1}{2\mu_0}\left|B_x(j+\frac{1}{2},k+\frac{1}{2})\right|^2\Big]\Delta y\Delta z \tag{8.4.1}$$

$$(j = 1,2,\cdots N_y;\ k = 1,2,\cdots N_z).$$

The initial value for electromagnetic wave energy is negligibly small as compared with the value at saturation W_s. Hence W_s can be regarded as the electromagnetic wave energy obtained from the electron beam. On the other hand, the total kinetic energy carried by superparticles for electrons contained in a small segment is given by

$$W_b = \sum_i m_e c^2(\gamma_i - 1), \tag{8.4.2}$$

where the summation is taken over all the superparticles for electrons contained in a small segment. Let the values of the kinetic energy at the initial time and saturation time be denoted by W_{b0} and W_{bs}, respectively. Then, we can confirm that the relation $W_{b0} - W_{bs} = W_s$ holds within a small error in the above simulation. Hence the efficiency of energy transfer from the electron beam to the electromagnetic wave η can be expressed as

$$\eta = \frac{W_s}{W_{b0}} = \frac{W_{b0} - W_{bs}}{W_{b0}}. \tag{8.4.3}$$

In the example treated in the above numerical simulation, we get $\eta = 2.69\%$.

8.5 Efficiency Enhancement by a Tapered Dielectric Grating

In this section, we show a design example of efficiency enhancement for energy transfer from the electron beam to the electromagnetic wave in a Cherenkov laser [8.10]. In the Cherenkov laser, the kinetic energy of a relativistic electron beam is converted into the electromagnetic wave energy through the active coupling between a relativistic electron beam and a slow electromagnetic wave propagated along a dielectric-loaded waveguide. Thus the growth of the electromagnetic wave in the Cherenkov laser causes the electron beam to lose kinetic energy, leading to the decrease in its average velocity or drift velocity. As a result, the electromagnetic wave gradually gets out of synchronism with the electron beam, since the former travels along the latter at a constant velocity. Hence, efficient transfer of energy from the electron beam to the electromagnetic wave is prevented. This difficulty can be overcome if the velocity of the electromagnetic wave is reduced by some means in accordance with the decrease in the average beam velocity, in order to keep synchronism between them. The phase velocity of the electromagnetic wave can be reduced in accordance with the decrease in the drift velocity of the electron beam, by gradually increasing the permittivity of the dielectric composing the waveguide along it. For a specific method to increase gradually the effective permittivity of a dielectric constituting the waveguide, we can use a waveguide loaded with a dielectric grating, in which either one of the grating parameters, or the slot width and the groove depth, is adiabatically decreased in accordance with the decrease in the drift velocity of the electron beam. In this section, with the aid of particle simulation, we demonstrate the validity of the scheme for efficiency enhancement presented above.

8.5.1 Effective Permittivity of a Dielectric Grating

The 2-D model of a Cherenkov laser loaded with a dielectric grating is illustrated in Fig. 8.10. To simplify the analysis, we replace a dielectric grating by a dielectric sheet with uniform thickness and effective relative permittivity ε_r'. The effective relative permittivity ε_r' is obtained by averaging the relative permittivities of the vacuum and dielectric regions included in a grating segment over one spatial period. Then, we get for the value of ε_r',

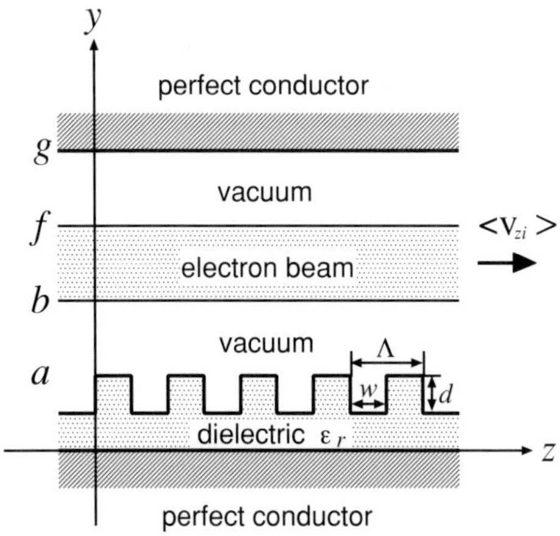

Fig. 8.10. 2-D model of the Cherenkov laser loaded with a dielectric grating

$$\varepsilon_r' = \frac{wd + \varepsilon_r[(\Lambda - w)d + (a-d)\Lambda]}{\Lambda a}, \tag{8.5.1}$$

where the parameters Λ, w, and d denote the period, the slot width and the groove depth of a dielectric grating, respectively. These parameters must be much smaller than the guide wavelength so that a dielectric grating can be replaced by an equivalent dielectric sheet with uniform thickness.

The Cherenkov instability occurs around the guide wavelength λ_g at which the phase velocity of the electromagnetic wave propagated along a parallel-plate waveguide loaded with a uniform dielectric sheet with thickness a and effective relative permittivity ε_r' becomes equal to the initial beam velocity v_0 or the synchronism condition is satisfied. In order to get the synchronism condition, let us first find the dispersion relation for the electromagnetic wave propagated along a new parallel-plate waveguide defined above, which is approximately equivalent to the original one. Referring to Sect. 5.3, we readily get for the dispersion relation,

$$\sqrt{\varepsilon_r'\left(\frac{\omega}{k_z}\right)^2 - 1}\,\tan\left[k_z a\sqrt{\varepsilon_r'\left(\frac{\omega}{k_z}\right)^2 - 1}\right]$$

$$= \varepsilon_r'\sqrt{1-\left(\frac{\omega}{ck_z}\right)^2}\,\tanh\left[k_z(g-a)\sqrt{1-\left(\frac{\omega}{ck_z}\right)^2}\right]. \tag{8.5.2}$$

The synchronism condition can be found by setting $\omega / k_z = v_0$ and $k_z = 2\pi / \lambda_g$ in (8.5.2) as follows:

$$\sqrt{\varepsilon_r' \beta_0^2 - 1} \tan\left[\frac{2\pi a}{\lambda_g}\sqrt{\varepsilon_r' \beta_0^2 - 1}\right]$$

$$= \varepsilon_r' \sqrt{1 - \beta_0^2} \tanh\left[\frac{2\pi(g-a)}{\lambda_g}\sqrt{1 - \beta_0^2}\right], \tag{8.5.3}$$

from which the guide wavelength is obtained. The more accurate value of the guide wavelength can be found by solving the exact boundary-value problem involving a relativistic electron beam.

8.5.2 Dependence of the Growth Characteristics on the Grating Parameters

Let us investigate how the growth characteristics for the electromagnetic wave depend on the grating parameters, or the slot width and the groove depth, of a dielectric grating. For this purpose, we calculate the average power P for the electromagnetic wave according to

$$P = -\frac{\Delta y}{\mu_0 N_z} \sum_{k=1}^{N_z} \sum_{j=0}^{N_y - 1} E_y\left(j + \frac{1}{2}, k\right) B_x\left(j + \frac{1}{2}, k\right), \tag{8.5.4}$$

which represents the electromagnetic power per unit length in the x direction averaged over one beam segment. The various dimensions used in the following simulation are listed in Table 8.3. The temporal growth of the electromagnetic power is shown in Fig. 8.11 when the groove depth is changed as a parameter, and in Fig. 8.12 when the slot width is changed as a parameter. As seen from Figs. 8.11 and 8.12, we get the less power and the higher frequency for the electromagnetic wave as the groove depth is deepened or the slot width is widened. On the other hand, we show in Fig. 8.13 the temporal growth of the electromagnetic power when the permittivity of a dielectric sheet with uniform thickness is changed as a parameter. From Fig. 8.13, we see that we get the less power and the higher frequency for the electromagnetic wave as the permittivity decreases. Comparing the results in Figs. 8.11 and 8.12 with those in Fig. 8.13, we find that varying the groove depth or the slot width in a dielectric grating with constant permittivity has the same effect on the growth characteristics for the electromagnetic wave as varying the permittivity of a dielectric sheet with constant thickness. In the above discussion, we have compared the growth characteristics when the grating parameters or the grating permittivity is changed as a parameter, keeping constant the initial beam voltage or the initial beam velocity. We can show a similar correspondence between varying the groove depth or the slot width in a dielectric grating with constant permittivity and varying the permittivity of a dielectric sheet with constant thickness, also when the frequency of the electromagnetic wave is kept invariant. From the above discussion, we can conclude that the effective permittivity of a dielectric grating can be adiabatically changed by gradually decreasing groove depths or slot widths in a dielectric grating with constant permittivity.

Table 8.3. Values of various parameters used in Sect. 8.5

Relative Permittivity of Dielectric ε_r	2.12
Thickness of Dielectric Grating a	0.5 [mm]
Period of Dielectric Grating Λ	0.25 [mm]
Beam Thickness $(f - b)$	0.125 [mm]
Beam-Grating Gap $(b - a)$	0.5 [mm]
Separation between Parallel Plates g	2.0 [mm]
Plasma Frequency for Electrons $\omega_p a/c$	0.01
Length of Beam Segment L	
(One Guide Wavelength λ_g)	2.0 [mm]
Number of Superparticles for Electrons N	1024
Number of Spatial Grids in the y Direction N_y	128
Number of Spatial Grids in the z Direction N_z	128
Spatial Grid Interval $\Delta y = \Delta z$	0.0156 [mm]
Time Interval for One Step $c\Delta t$	0.008994[mm]

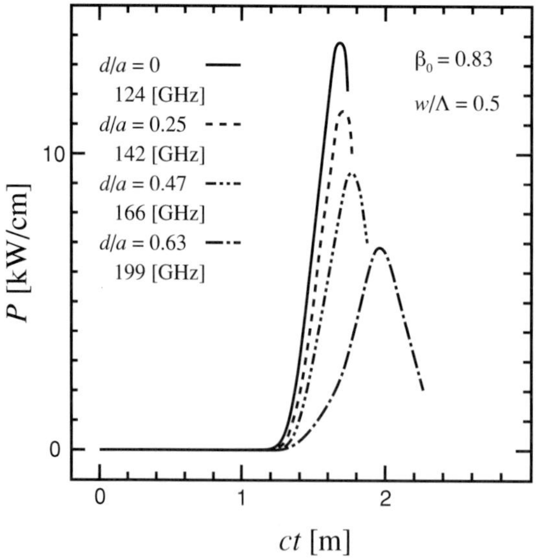

Fig. 8.11. Variation of the growth characteristics with the groove depth of a dielectric grating changed as a parameter

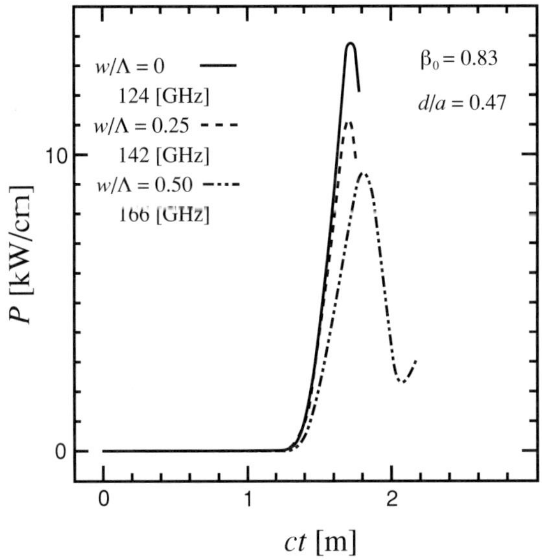

Fig. 8.12. Variation of the growth characteristics with the slot width of a dielectric grating changed as a parameter

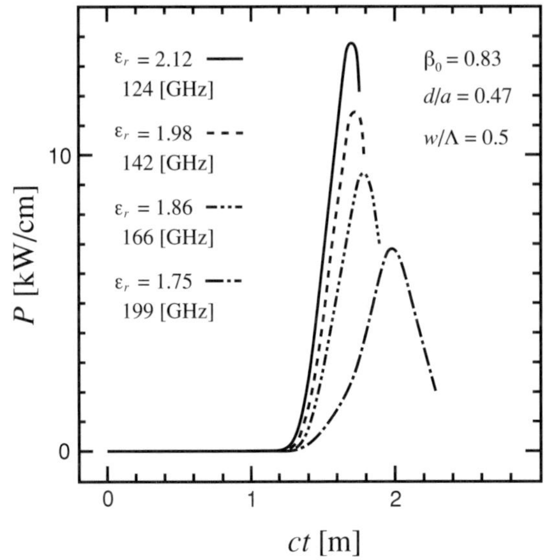

Fig. 8.13. Variation of the growth characteristics with the relative permittivity of a dielectric grating changed as a parameter

8.5.3 Efficiency Enhancement

As described in the beginning of this section, the efficiency of energy transfer from the electron beam to the electromagnetic wave in the Cherenkov laser can be

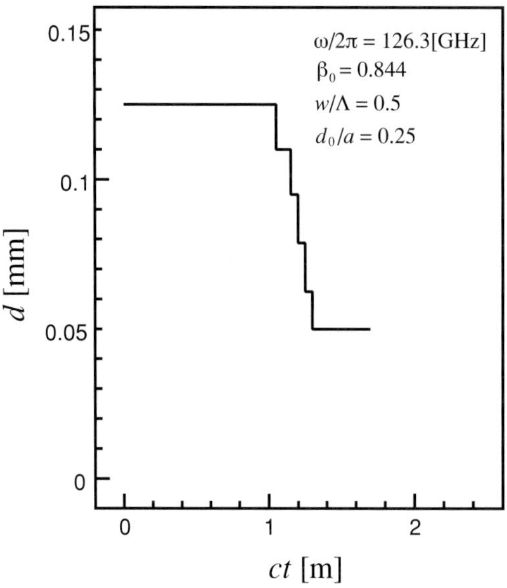

Fig. 8.14. Optimum variation of the groove depth of a dielectric grating

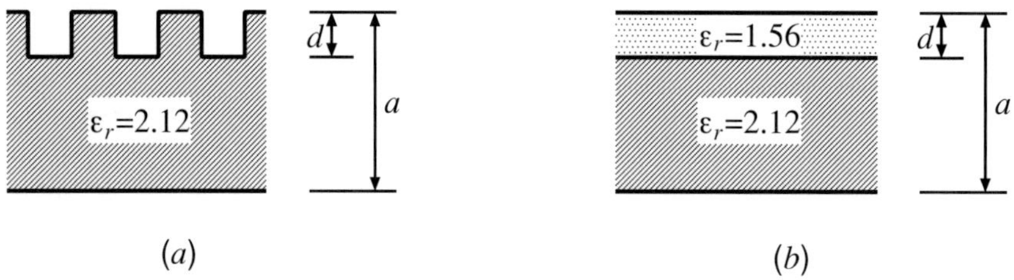

Fig. 8.15. Replacement of a dielectric grating by a couple of uniform dielectric sheets

enhanced by gradually increasing the permittivity of a dielectric sheet with constant thickness along the waveguide, in accordance with the decrease in the average velocity of the electron beam. As demonstrated in the foregoing discussion, the effective permittivity of a dielectric grating can be controlled by varying either one of the grating parameters, or the slot width and the groove depth. Hence the efficiency of energy transfer can be enhanced by using a properly tapered dielectric grating. Since we can get similar results for variable slot widths or variable groove depths, we demonstrate the efficiency enhancement for variable groove depths.

For a given set of various parameters, the optimum variation of the grating parameter or the groove depth with the interaction time t is shown in Fig. 8.14, in

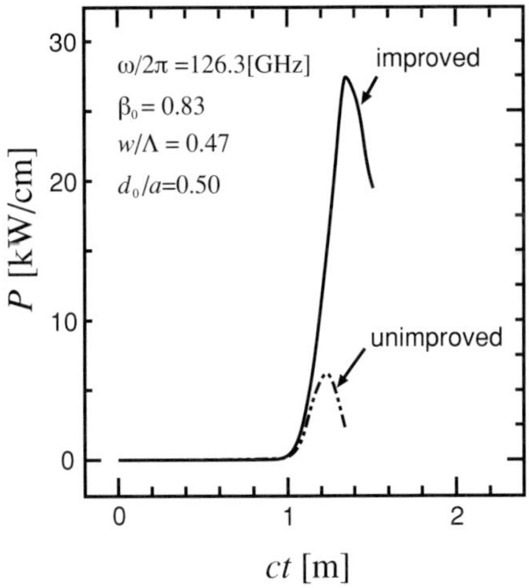

Fig. 8. 16. Efficiency en-hancement by a tapered dielectric grating

which the groove depth d is changed in such a way that the synchronism between the phase velocity of the electromagnetic wave and the average velocity of the electron beam can be kept as the electromagnetic wave grows. In the simulation of Fig. 8.14, we have used 0.125mm for the initial value of the groove depth d_0, and 0.25mm for the value of the beam-grating gap. To determine the optimum groove depth of a dielectric grating for a particular average velocity of the electron beam, we replaced a dielectric grating by two uniform dielectric sheets with constant thicknesses, as shown in Fig. 8.15. The upper sheet in Fig. 8.15 has the thickness d and the relative permittivity1.56 averaged over the vacuum and dielectric regions, while the lower sheet has the thickness $a - d$ and the relative permittivity 2.12. The above replacement is justified for groove depths much smaller than the guide wavelength of the electromagnetic wave. For the optimum variation of the groove depths given in Fig. 8.14, we illustrate in Fig. 8.16 the variation of the electromagnetic power P extracted from the electron beam per unit length in the x direction. In order to clarify the improvement in the growth characteristics for the electromagnetic wave, we also show in Fig. 8.16 those for a dielectric grating with constant parameters. From Fig. 8.16, we see that the growth characteristics for the electromagnetic wave can be greatly improved by properly tapering the groove depths of a dielectric grating. For the case shown in Fig. 8.16, the efficiency of energy transfer is equal to 1.5 % for a dielectric grating with constant parameters while it is enhanced to 6.85 % for a dielectric grating with properly tapered groove depths.

In the numerical simulation in this section, a segment of the electron beam is represented by 8 rows of superparticles, each of which has the sum of 128 superparticles for electrons. At $t = 0$, the particles in each row are assumed to be equally

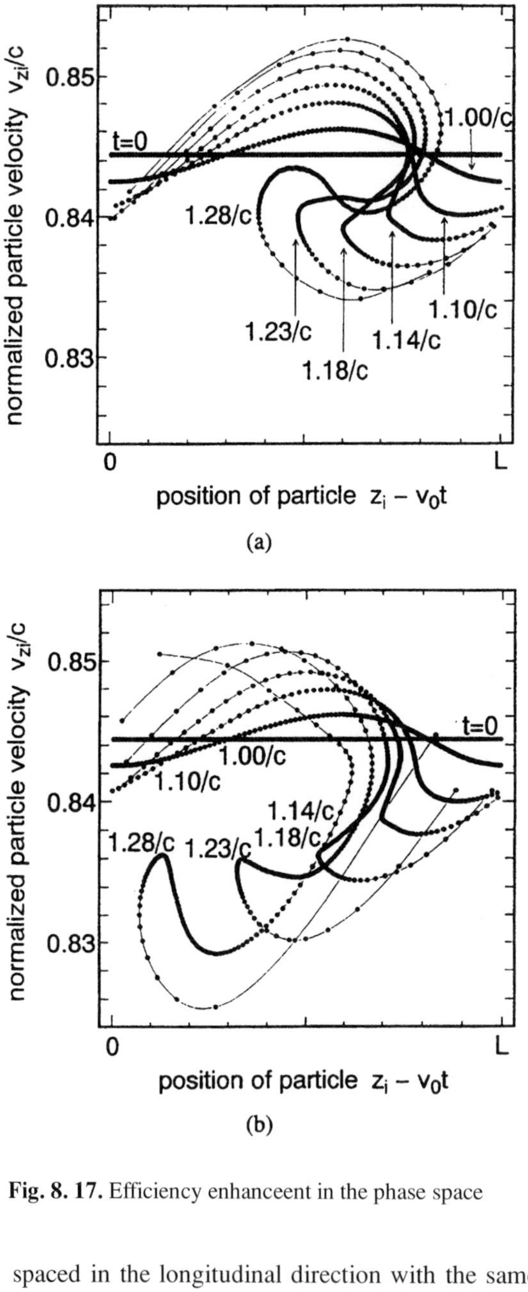

Fig. 8. 17. Efficiency enhanceent in the phase space

spaced in the longitudinal direction with the same velocity v_0. To get some idea of how the positions and velocities of the particles in a row of the beam segment are changed as a result of their interaction with the electromagnetic wave, we temporally follow in Fig. 8.17 the positions and velocities of the particles in that row of the beam segment which is nearest the surface of the dielectric grating. For reference, we show in Fig. 8.17 (*a*) the phase-space plot for a dielectric grating with constant parameters. In Fig. 8.17 (*b*), on the other hand, the phase-space plot for a dielectric grating with variable groove depths is shown. As is evident from Fig. 8.17 (*b*), the bunched particles remain trapped in the decelerating phase of the electric field for longer time and shift to lower velocities as the electromagnetic wave slows down with decreasing groove depths of a dielectric grating. We also see from Fig. 8.17 (*b*) that the bunched particles shift to the left with decreasing phase velocity of the electromagnetic wave, since the framework observing particles is moving with the initial beam velocity v_0. Comparing the results shown in Figs. 8.17 (*a*) and 8.17 (*b*), we find that the larger amount of energy can be extracted from the electron beam for a dielectric grating with decreasing groove depths than for a dielectric grating with constant parameters.

References

Chapter 1

1.1 J.D. Jackson: *Classical Electrodynamics*, 2nd edn. (Wiley, New York 1975)
1.2 W.K.H. Panofsky, M. Phillips: *Classical Electricity and Magnetism*, 2nd edn. (Addison-Wesley, Massachusetts 1962)
1.3 J.A. Stratton: *Electromagnetic Theory* (McGraw-Hill, New York 1941)
1.4 J.A. Kong: *Electromagnetic Wave Theory* (Wiley, New York 1986)
1.5 R.E. Collin: *Foundations for Microwave Engineering* (McGraw-Hill, New York 1966)
1.6 S. Ramo, J.R. Whinnery, T. Van Duzer: *Fields and Waves in Communication Electronics* (Wiley, New York 1984)
1.7 L.D. Landau, E.M. Lifshitz: *Electrodynamics of Continuous Media* (Pergamon, Oxford 1960)
1.8 V.L. Ginzburg: *The Propagation of Electromagnetic Waves in Plasmas* (Pergamon, Oxford 1964)
1.9 K.C. Yeh, C.H. Liu: *Theory of Ionospheric Waves* (Academic, New York 1972)

Chapter 2

2.1 W. Pauli: *Theory of Relativity* (Pergamon, Oxford 1958)
2.2 C. Mϕller: *The Theory of Relativity* (Clarendon, Oxford 1972)
2.3 A. Sommerfeld: *Electrodynamics* (Academic, New York 1952)
2.4 J. Van Bladel: *Relativity and Engineering* (Springer, Berlin 1984)
2.5 A. Einstein: "Zur Elektrodynamik Bewegter Körper", Ann. Phys. **17**, 891 (1905)
2.6 H. Minkowski: Die Grundgleichungen für die Elektromagnetischen Vorgänge in Bewegten Körpern", Göttinger Nachrichten, 53 (1908)
2.7 C.T. Tai: "A Study of Electrodynamics of Moving Media", Proc. IEEE **52**, 685 (1964)
2.8 C.T. Tai: "A Present View on Electrodynamics of Moving Media", Radio Sci. **2**, 245 (1967)
2.9 T. Shiozawa: "Phenomenological and Electron-Theoretical Study of the Electrodynamics of Rotating Systems", Proc. IEEE **61**, 1694 (1973)
2.10 J.A. Kong: *Electromagnetic Wave Theory* (Wiley, New York 1986)
2.11 J.D. Jackson: *Classical Electrodynamics*, 2nd edn. (Wiley, New York 1975)
2.12 R.C. Costen, D. Adamson: "Three-Dimensional Derivation of the Electrodynamic Jump Conditions and Momentum-Energy Laws at a Moving Boundary", Proc. IEEE **53**, 1181 (1965)
2.13 A. Einstein: "Das Prinzip von der Erhaltung der Schwerpunktsbewegung und die Trägheit der Energie", Ann. Phys. **20**, 627 (1906)
2.14 A.P. French: *Special Relativity* (The MIT Introductory Physics Series) (MIT, Cambridge 1971)

Chapter 3

3.1 J.D. Jackson: *Classical Electrodynamics*, 1st edn. (Wiley, New York 1962)
3.2 W.K.H. Panofsky, M. Phillips: *Classical Electricity and Magnetism*, 2nd edn. (Addison-Wesley, Massachusetts 1962)
3.3 L.D. Landau, E.M. Lifshitz: *Classical Theory of Fields* (Pergamon, Oxford 1971)
3.4 H. Wiedemann: *Synchrotron Radiation* (Springer, Berlin 2003)
3.5 P.A. Cherenkov: "Visible Radiation Produced by Electrons Moving in a Medium with Velocities Exceeding that of Light", Phys. Rev. **52**, 378 (1937)
3.6 I.M. Frank, I.E. Tamm: "Coherent Visible Radiation of Fast Electrons Passing through Matter", Compt. Rend. de l'Acad. Sci. U.R.S.S. **14**, 109 (1937)
3.7 L.D. Landau, E.M. Lifshitz: *Electrodynamics of Continuous Media* (Pergamon, Oxford 1960) p. 357
3.8 L.I. Schiff: *Quantum Mechanics*, 2nd edn. (McGraw-Hill, New York 1955) p. 267
3.9 J.A. Kong: *Electromagnetic Wave Theory* (Wiley, New York 1986) p. 220
3.10 L. Schächter: *Beam-Wave Interaction in Periodic and Quasi-Periodic Structures* (Springer, Berlin 1996) p. 49

Chapter 4

4.1 R.B. Miller: *An Introduction to the Physics of Intense Charged Particle Beams* (Plenum, New York 1982)
4.2 J.R. Pierce: *Theory and Design of Electron Beams* (Van Nostrand, New York 1954)
4.3 R.E. Collin: *Foundations for Microwave Engineering* (McGraw-Hill, New York 1966)
4.4 T. Shiozawa: "Electrodynamics of Rotating Relativistic Electron Beams", Proc. IEEE **66**, 638 (1978)
4.5 L. Schächter: *Beam-Wave Interaction in Periodic and Quasi-Periodic Structures* (Springer, Berlin 1996)
4.6 C.L. Hemenway, R.W. Henry, M. Caulton: *Physical Eelectronics*, 2nd edn. (Wiley, New York 1967)
4.7 W.C. Hahn: "Small Signal Theory of Velocity-Modulated Electron Beams", Gen. Elec. Rev. **42**, 258 (1939)
4.8 S. Ramo: "Space-Charge and Field Waves in an Electron Beam", Phys. Rev. **56**, 276 (1939)
4.9 P.A. Sturrock: "A Variational Principle and an Energy Theorem for Small-Amplitude Disturbances of Electron Beams and of Electron-Ion Plasmas", Ann. Phys. **4**, 306 (1958)
4.10 A. Bers, P. Penfield, Jr.: "Conservation Principles for Plasmas and Relativistic Electron Beams", IRE Trans. Electron Devices **ED-9**, 12 (1962)
4.11 D.L. Bobroff, H.A. Haus, J.W. Klüver: "On the Small Signal Power Theorem of Electron Beams", J. Appl. Phys. **33**, 2932 (1962)
4.12 J. A. Kong: "Theorems of Bianisotropic Media", Proc. IEEE **60**, 1036 (1972)
4.13 W.P. Allis, S.J. Bucksbaum, A. Bers: *Waves in Anisotropic Plasmas* (MIT, Cambridge1962) Chaps. 7, 8
4.14 T. Musha, M. Agu: "Energy and Power Flow of a Wave in a Moving Dispersive Medium", J. Phys. Soc. Japan **26**, 541 (1969)
4.15 M. Agu, Y. Daido, T. Takagi: "On the Energy and Momentum Conservation Laws for Linearized Electromagnetic Fields in a Dispersive Medium", J. Phys. Soc. Japan **33**, 519 (1972)
4.16 P.A. Sturrock: "In What Sense Do Slow Waves Carry Negative Energy?", J. Appl. Phys. **31**, 2052 (1960)

4.17 R.J. Briggs: Transformation of Small-Signal Energy and Momentum of Waves", J. Appl. Phys. **35**, 3268 (1964)
4.18 J.R. Pierce: "Momentum and Energy of Waves", J. Appl. Phys. **32**, 2580 (1961)
4.19 T. Musha: "Electrodynamics of Anisotropic Media with Space and Time Dispersion", Proc. IEEE **60**, 1475 (1972)
4.20 H. Kogelnik, H.P. Weber: "Rays, Stored Energy, and Power Flow in Dielectric Waveguides", J. Opt. Soc. Amer. **64**, 174 (1974)
4.21 H.A. Haus, H. Kogelnik: "Electromagnetic Momentum and Momentum Flow in Dielectric Waveguides", J. Opt. Soc. Amer. **66**, 320 (1976)

Chapter 5

5.1 J.R. Pierce: *Traveling-Wave Tubes* (Van Nostrand, New York 1950)
5.2 R.E. Collin: *Foundations for Microwave Engineering* (McGraw-Hill, New York 1966)
5.3 K.L. Felch, K.O. Besby, R.W. Layman, D. Kapilow, J.E. Walsh: "Cerenkov Radiation in Dielectric-Lined Waveguides", Appl. Phys. Lett. **38**, 601(1981)
5.4 J.E. Walsh, J.B. Murphy: "Tunable Cerenkov Lasers", IEEE J. Quant. Electron. **QE-18**, 1259 (1982)
5.5 E.P. Gerate, J.E. Walsh: "The Cerenkov Maser at Millimeter Wavelengths", IEEE Trans. Plasma Sci. **PS-13**, 524 (1985)
5.6 M. Shoucri: "The Excitation of Microwaves by a Relativistic Electron Beam in a Dielectric-Lined Waveguide", Phys. Fluids **26**, 2271 (1983)
5.7 V.K. Tripathi: "Excitation of Electromagnetic Waves by an Axial Electron Beam in a Slow Wave Structure", J. Appl. Phys. **56**, 1953 (1984)
5.8 T. Shiozawa, H. Kondo: "Mode Analysis of an Open-Boundary Cherenkov Laser in the Collective Regime", IEEE J. Quant. Electron. **QE-23**, 1633 (1987)
5.9 L. Schächter: *Beam-Wave Interaction in Periodic and Quasi-Periodic Structures* (Springer, Berlin 1996)
5.10 W.H. Louisell: *Coupled Mode and Parametric Electronics* (Wiley, New York 1960)
5.11 K. Horinouchi, T. Shiozawa: "Analysis of Dynamic Behavior of an Open-Boundary Cherenkov Laser", Electron. & Commun. Japan, Part 2, **77**, 12 (1994)
5.12 A. Yariv, C-C. Shih: "Amplification of Radiation by Relativistic Electrons in Spatially Periodic Optical Waveguides", Opt. Commun. **24**, 233 (1978)
5.13 A. Sommerfeld: *Mechanics* (Academic, New York 1952) Chap. 3

Chapter 6

6.1 T.C. Marshall: *Free-Electron Lasers* (Macmillan, New York 1985)
6.2 A. Yariv: *Quatum Electronics*, 3rd edn. (Wiley, New York 1989) Chap. 13
6.3 L. Schächter: *Beam-Wave Interaction in Periodic and Quasi-Periodic Structures* (Springer, Berlin 1996) Chap. 7
6.4 H. Motz: "Applications of the Radiation from Fast Electron Beams", J. Appl. Phys. **22**, 527 (1951)
6.5 R.M. Phillips: "The Ubitron, a High Power Traveling Wave Tube Based on a Periodic Beam Interaction in Unloaded Waveguide", IRE Trans. Elec. Dev. **7**, 231 (1960)
6.6 J.M.J. Madey: "Stimulated Emission of Bremsstrahrung in a Periodic Magnetic Field", J. Appl. Phys. **42**, 1906 (1971)
6.7 L.R. Elias, W.M. Fairbank, J.M.J. Madey, H.A. Schwettman, T.I. Smith: "Observation of Stimulated Emission of Radiation by Relativistic Electrons in a Spatially Periodic Transverse Magnetic Field", Phys. Rev. Lett. **36**, 717 (1976)

6.8 D.A.G. Deacon, L.R. Elias, J.M.J. Madey, G.J. Ramian, H.A. Schwettman, T.I. Smith: "First Operation of a Free-Electron Laser", Phys. Rev. Lett. **38**, 892 (1977)
6.9 W.B. Colson: "One Body Analysis of Free-Electron Lasers", in *Physics of Quantum Electronics*, Vol. 5, ed. by S.F. Jacobs, M. Sargent III, M.O. Scully (Addison-Wesley, Massachusetts 1978)
6.10 P.L. Morton: "Relationship of FEL Physics to Accelerator Physics", in *Physics of Quantum Electronics*, Vol. 8, ed. by S.F. Jacobs, G.T. Moore, H.S. Pilloff, M. Sargent III, M.O. Scully, R. Spitzer (Addison-Wesley, Massachusetts 1982)
6.11 T.J. Orzechowski, B.R. Anderson, J.C. Clark, W.M. Fauley, A.C. Paul, D. Prosnitz, E.T. Scharlemann, S.M. Yarema: "High-Efficiency Extraction of Microwave Radiation from a Tapered-Wiggler Free-Electron Laser", Phys. Rev. Lett. **57**, 2172 (1986)
6.12 N.M. Kroll, P.L. Morton, M.N. Rosenbluth: "Free-Electron Lasers with Variable Parameter Wigglers", IEEE J. Quant. Electron. **QE-17**, 1436 (1981)

Chapter 7

7.1 T.C. Marshall: *Free-Electron Lasers* (Macmillan, New York 1985)
7.2 L. Schächter: *Beam-Wave Interaction in Periodic and Quasi-Periodic Structures* (Springer, Berlin 1996) Chap. 7
7.3 R.H. Pantell, G. Soncini, H.E. Puthoff: "Stimulated Photon-Electron Scattering", IEEE J. Quant. Electron. **QE-4**, 905 (1968)
7.4 P. Sprangle, V.L. Granatstein, L. Baker: "Stimulated Collective Scattering from a Magnetized Relativistic Electron Beam", Phys. Rev. A **12**, 1697 (1975)
7.5 T. Kwan, J.M. Dawson, A.T. Lin: "Free Electron Laser", Phys. Fluids **20**, 581 (1977)
7.6 V.L. Granatstein, P. Sprangle: "Mechanisms for Coherent Scattering of Electromagnetic Waves from Relativistic Electron Beams", IEEE Trans. Microwave Theory and Tech. **MTT-25**, 545 (1977)
7.7 A. Hasegawa: "Free Electron Laser", Bell Syst. Tech. J. **57**, 3069 (1978)
7.8 T. Shiozawa: "Stimulated Raman Scattering by a Relativistic Electron Beam – Analysis of a Free-Electron Laser", Electron. & Commun. Japan **63-C**, 97 (1980)
7.9 T. Shiozawa: "A General Theory of the Raman-Type Free-Electron Laser", J. Appl. Phys. **54**, 3712 (1983)
7.10 T. Shiozawa, T. Nakashima: "Two-Dimensional Mode Analysis of the Raman-Type Free-Electron Laser", J. Appl. Phys. **54**, 3712 (1983)
7.11 J.A Armstrong, N. Bloembergen, J. Ducuing, P.S. Pershan: "Interaction between Light Waves in a Nonlinear Dielectric", Phys. Rev. **127**, 1918 (1962)
7.12 Y.R. Shen: *The Principles of Nonlinear Optics* (Wiley, New York 1984)
7.13 J.C. Slater: *Microwave Electronics* (Wiley, New York 1950)

Chapter 8

8.1 R.W. Hockney, J.W. Eastwood: *Computer Simulation Using Particles* (McGraw-Hill, New York 1981)
8.2 J.M. Dawson, A.T. Lin: "Particle Simulations", in *Basic Plasma Physics* ed. by A.A. Galeev, R.N. Sudan (Elsevier, Amsterdam 1984)
8.3 C.K. Birdsall, A.B. Langdon: *Plasma Physics via Computer Simulation* (McGraw-Hill, New York 1985)
8.4 K.S. Yee: "Numerical Solution of Initial Boundary-Value Problems Involving Maxwell's Equations in Isotropic Media", IEEE Trans. Antennas & Propagat. **14**, 302 (1966)
8.5 A. Taflove: *Computational Electrodynamics –The Finite-Difference Time-Domain*

Method (Artech House, Massachusetts 1995)
8.6 T.D. Pointon, J.S. De Groot: "Particle Simulation of Plasma and Dielectric Cerenkov Masers", Phys. Fluids **31**, 908 (1988)
8.7 L. Schächter: "Cerenkov Traveling-Wave Tube with a Spatially Varying Dielectric Coefficient", Phys. Rev. A **43**, 3785 (1991)
8.8 K. Horinouchi, M. Sanda, H. Takahashi, T. Shiozawa: Analysis of a Cherenkov Laser via Particle Simulation" (in Japanese), IEICE Trans. **J-78-C-I**, 1 (1995)
8.9 T. Shiozawa, T. Yoshitake: "Efficiency Enhancement in a Cherenkov Laser Loaded with a Kerr-like Medium", IEEE J. Quant. Electron. **QE-31**, 539 (1995)
8.10 T. Shiozawa, H. Takahashi, Y. Kimura: "Nonlinear Saturation and Efficiecy Enhancement in a Cherenkov Laser Using a Dielectric Grating", IEEE J. Quant. Electron. **QE 32**, 2037 (1996)
8.11 A. Hirata, S. Hirosaka, T. Shiozawa: "Effectiveness and Limitation of the Periodic Boundary Approximation in the Analysis of Single-Pass Electron Beam Devices", IEEE Trans. Plasma Sci. **30**, 1292 (2002)

Index

Printing: Saladruck, Berlin
Binding: Stein+Lehmann, Berlin